Flying into Danger

Patrick Forman was educated at Loretto School
and Cambridge and Edinburgh universities. He
first became interested in flying when he served in
the war as a platoon commander with the 1st
Airborne Division (Gliders). He started flying as a
private pilot in 1969 and now has about 1,000
command hours behind him. After practising as a
solicitor he was the *Sunday Times* air
correspondent for ten years. He now contributes
to press and television as an aviation consultant,
and is a member of the Air Safety Group, a
Parliamentary advisory committee. He is married
and lives in Cambridge.

PATRICK FORMAN

Flying into Danger

The Hidden Facts about Air Safety

Mandarin

A Mandarin Paperback
FLYING INTO DANGER

First published in Great Britain 1990
by William Heinemann Ltd
This edition published 1991
by Mandarin Paperbacks
Michelin House, 81 Fulham Road, London SW3 6RB

Mandarin is an imprint of the Octopus Publishing Group,
a division of Reed International Books Limited

Copyright © Patrick Forman 1990

A CIP catalogue record for this title
is available from the British Library
ISBN 0 7493 0922 9

Printed and bound in Great Britain
by Cox & Wyman Ltd, Reading, Berks

Contents

Acknowledgments

I wish to acknowledge my debt to the many people who have helped me in producing this book. Among the foremost of these is Jim Burnett, Chairman of the United States National Transportation Safety Board, who continued to provide a flow of necessary documentation. I owe much to long-standing liaisons with Michael Young and his staff at the UK Civil Aviation Authority's Public Relations Department; to Mark Young and his colleagues of the British Airline Pilots' Association who also allowed me to quote material from BALPA's magazine *The Log*; to Michael Ramsden, editor in chief of *Flight International* and its editor David Mason and staff; to chief inspectors and colleagues at the Air Accident Investigation Branch of the Department of Transport at Farnborough, England; and to Roger Green of the Human Factors Division, Royal Air Force Institute of Aviation Medicine.

My thanks also go to Captain Eric Moody of British Airways for his experiences over Jakarta in the 1982 volcanic eruption. Special gratitude is due to Lufthansa German Airlines not only for many instructive trips on flight decks, but also to Captain Heino Caesar its General Manager of Flight Operations Inspections and Safety Pilot for extracts from his 1989 study of jet-loss statistics; and to Herr Ugo Wargenau, Engineering Director, for providing research material on cabin fire-proofing.

As a member of the Parliamentary Air Safety Group, I have benefited greatly over the years from colleagues, in particular its past chairman Norman Tebbit, its present chairman Captain Eric Pritchard, and its secretary Frank Taylor of the Cranfield Institute of Technology. I would also like to broadcast by appreciation to all those pilots and others who, while preferring to remain anonymous, have been a constant inspiration in the quest for safety in the air. One who can be named is Eric Thurston whose hands-on airmanship taught me more than any text book could do. And in retrospect, the book owes its origins to the encouragement and enthusiasm of Harold Evans while he was my editor at the *Sunday Times*.

Finally, nothing would have been possible without the editorial patience of my wife Sarah, who has been my literary co-pilot throughout what might be called a most extended operation.

Patrick Forman
January 1990

Introduction

At nine o'clock on a snowy April morning, the party of young mothers bent on a day's shopping trip from England to Switzerland excitedly gathered their belongings as their plane prepared to land. Those sitting near the windows caught a glimpse of the city of Basle through the snow flurries. Some had never flown before. But the holiday spirit is infectious and should have suppressed any twinges of apprehension they felt as the ground grew closer.

It was no ordinary treat. The Axbridge Ladies' Guild had arranged a very special outing for friends and neighbours from a cluster of five little English West Country villages through a travel agent in nearby Bristol. The hamlets of Cheddar, Wrington, Congresbury and Yatton lie near the slopes of the Mendip hills, all of them within eight miles of Axbridge's Gothic church and thirteenth-century royal hunting lodge. Bristol airport is conveniently close across the valley of the Yea trout-stream.

On the day of the flight, husbands and children had been left at home. It was Mums' Day Out and nothing, not even the weather, was going to dismay their high hopes of a spree in the Foreign shops. What they did not know was that, on the other side of the cabin door that screened them from the flight crew, their two pilots were lost. And what finally sealed their fate was that the pilots did not know this either.

Their plane had just overflown the city of Basle and was heading for a spot on the snow-clad Jura mountains, where a stone memorial now stands with 108 of their names engraved upon it. Only 35 of the party that set out from Bristol that morning returned. But as the plane headed south towards the mountains they still had ten minutes to live – ten minutes which should have saved their lives.

Captain Anthony Noel Dorman was at the controls of the four-engined Vickers Vanguard turboprop. He was a 35-year-old Canadian ex-military flier. On his right was his co-pilot Captain Ivor

Terry, at 47 a more experienced British airman, but it was company practice for pilots to take turns at command and this trip – Flight 435 – fell to Dorman. As he had descended towards Basle, he had wrongly identified two vital radio beacons on the ground and missed another. This led him to start his approach not four miles to the north of the airport but directly over it, thus flying south towards the mountains in search of the runway that lay behind him.

Let us not trouble here with the technical side of Dorman's lapses in airmanship. It is enough to say that his abysmal wanderings led to both pilots' becoming hopelessly disorientated, as can so easily happen when the orderly sequence of navigation safety checks is broken under stress of 'blind' flying, which is done by relying entirely on instruments.

As co-pilot, Terry had been operating the radio and making the usual exchanges with the airport controllers. Shortly after the plane passed over the airport – unseen by the pilots glued to their instruments – he must have sensed that the approach was going badly wrong and took over control. It was Dorman's voice that now radioed to the airport – '435 overshooting'. Terry gave the engines power and climbed to the right, intending to start a new approach from scratch. But in the navigational mess that he had inherited from Dorman, neither of them realized that they were far south of the airport. Terry began a second and more accurate approach, but he based it on the wrong beacon sited on the runway and not the correct initial 'fix' four miles to the north.

Once again the plane passed the airport and started its descent towards the mountains. An amazed official at work in the city observatory saw the red-painted tail of the British Vanguard clear the building by only 150 feet and rushed to telephone the airport – 'It is snowing very heavily, if it remains like this it will crash in the mountains.' Three minutes later Zürich radar station saw an aircraft far south of Basle and warned the airport traffic controllers. They radioed Flight 435: 'Are you sure you are over BN [the correct beacon to the north]?' Dorman wrongly confirmed that they were, adding a routine report of his altitude. They were his last words.

Basle then replied: 'No, I think you are south of the field, you are on the south of the field . . .'

The crash came ten miles to the south of Basle in the peaks of the Jura mountains. Dorman and Terry were both killed. Two of the four air hostesses survived. In all, 108 of the souls who had set out from Somerset two hours earlier were lost.

In the cool light of day, accident analysts probably would not rate this air disaster as especially memorable, or one that contributed great lessons to the future safety of air travel. It happened as long ago as 1973 and – apart from the 55 children who became motherless on that day and the small community that bred them – it is largely lost in public memory. Yet among the 200 major air disasters that the world has had to suffer between 1973 and 1988, to me it remains as the most poignant. At the time, the media pounced so predictably on the human tragedy of a hundred and more friends and neighbours perishing in a common calamity that left so many young families without mothers. But they were soon forgotten.

I believe that even hardened professionals, who must live with the aftermath of air disasters, would own to a lasting and bitter underswell of sadness that comes with the job. The needless loss of lives – as it was in the Swiss aircrash – can fan the sadness into smouldering anger, but that is a profitless emotion unless harnessed to a resolve for action to prevent the predictable and avoidable causes of such 'accidents'.

The *aftermath* of that frightful day is more eloquent in the search for safety than the bare history of the flight. Swiss accident investigators claimed that Dorman had faked his pilot's licence by making false declarations about his qualifications. Before his civil flying career, he had been suspended from the Canadian Air Force for lack of ability. He had failed no less than eight times before finally passing his British instrument flying ratings – the cornerstone of airline piloting competence. (He would not have slipped through under the tougher regulations that were introduced as a result of the crash.) The reconstruction of his flight path after the crash – a serpentine tracing of frightening incompetence – and his neglect of a series of vital checks that should have warned him of the mounting danger more than endorse the Canadian verdict on his talents as a pilot.

But there were mitigations. Few crashes are due to a single cause, and this was no exception. The investigators found that repairs to one vital landing instrument had been botched and another revealed serious and uncorrected errors. This, they found, made the crew's navigational work more difficult. Neither the type of radio beacons nor the radar at Basle were up to modern standards. But good airmanship is meant to recognize and respond safely to such deficiencies.

Are there still lessons to be learnt from this dusty file of facts from the archives? A half-forgotten cautionary tale about a gimcrack provincial operation that ended in disaster almost a generation ago may seem an oblique approach to a study of air safety based on the many and much worse aircrashes that the world has had to suffer since then. My answer is that there is still a common factor that links them all and lives on to trouble us. The memory of the Axminster ladies throws into high relief the innocence and faith of air travellers – an innocence that is fostered within the airline industry, valued beyond price by the tourist and travel trade, and bolstered by the inveterate secrecy of officialdom.

On the day when the chairman of the Ladies' Guild and her friends arranged their outing with Unicorn Travel in Bristol, any idea that it might be a risky venture would have been far from their minds. Mrs Sylvia Cole was one of them. She survived the crash but her daughter was killed. She said of the flight later: 'It didn't enter my head that there was going to be a crash,' and even when the plane was floundering through the snow storm she had 'just hoped and prayed it would make a safe landing'. Before the event neither she nor her companions had had the slightest qualms. They were flying with a British airline, and the world knows that they are among the safest. But the flaws in man and machine already existed and lay dormant and hidden under a panoply of misinformation about the safety of travelling by air.

They shared, as members of the general public, all those received notions about the reliability, comfort and safety of airline travel contained in the glamorous liturgy of the industry. Modern air travel, as we the public are constantly assured, is far safer than motoring: we are told that it is more dangerous to cross the road than fly around the world, and that (according to the statistics of the day) the chance of being killed is greater than a million or so to one.

The figures are perfectly true – in one sense. They are based on world *average* statistics. Other things being equal, they reflect the chances of the *average* passenger on the *average* airliner and crew flying an *average* route. But the reality was that on that Tuesday morning, other things were not equal, and the fact is that no single

4

flight is an average flight. In the case of Flight 435, there were lies hidden in both the machine and in the man who flew it. The forces of secrecy, of not asking too many questions, of little fibs and fudged records, good advertising and smart sales talk, had all served to keep the image of airline safety intact.

Dorman had cheated his way through a mountain of safety checks on the qualification of airline pilots. Nobody, except himself, knew that he was as incompetent as he was. Nobody – neither the aviation authorities nor the airline – had bothered to find out. 'God only knows how he slipped through the net so long,' commented the chairman of his pilots' association after the event. 'I've never heard of another pilot failing his instrument rating test eight times. God only knows how he convinced the examiners at the airline that he was fit to fly.' Newspaper headlines shrieked: 'THE CHEAT – WHY WASN'T HE STOPPED?'

Unseen in the bowels of the old Vanguard turboprop lay the shoddily soldered navigation equipment and the faultily adjusted landing instrument, both of which later cast doubt on the actual legality of the flight. Pilots' complaints about these faults had earlier been ignored and the airline's records were inaccurate. On top of all this, Captain Dorman stepped up into the commander's seat just as worsening blizzards were beginning to sweep the runway at Basle.

The ladies' chances of survival were no longer a million to one.

The first aim of this book is to offer airline passengers an insight into the affairs that govern their fate. It seeks to provide some answer to the innumerable questions about flying that are put to those concerned with aviation, a demand that has kept pace with the vast growth in airline transport. At the core of this widespread interest there lies a profound concern about the safety of air travel.

More importantly, it is argued that the more people know about it, the safer it will become: time and again the record shows that public pressure –whether it is reflected in the media or by other means – is the mainspring of progress towards higher safety standards. However dedicated the people responsible for our safety may be – the government regulators, the airlines and the planemakers – in the last resort it is the public's monitoring eye that keeps them up to scratch. Confidences behind closed doors spell trouble and, as will be seen, they are too often a prelude to disaster. If this book helps to reaffirm the right of air passengers to know about the affairs that affect their safety, it will have done its job.

The fear of flying

Nobody denies that airline travel is remarkably safe and that it is getting safer. Why is it then, that so many of us are apprehensive about travelling by air when we know that it is far safer than making the journey by road. There is eminent support for those of us who think nothing of piling the family into the same car, yet prefer to split up the party into separate flights when it comes to a journey by air: the Queen, the Duke of Edinburgh and Prince Charles as heir-apparent follow the principle that only one may fly in the same aircraft, yet she and the Duke frequently travel in the same royal car.

On the available facts, this remarkable distrust of the aeroplane appears to be too persistent to dismiss as a mere human foible. The gap between what the safety statistics are telling us and what we actually feel about safety asks for an explanation, and I began to sense a missing dimension in the talk about airline safety. In studying the reports of accident inquiries, I soon became aware of a common factor that questioned the use of averages as a meaningful yardstick of the risks that face the passenger as an individual. It led me to the conclusion that those who are frightened by the prospect of flying may be wiser than they know. One of the aims of this book is to demonstrate that we are not being given all the facts needed to assess the real risks of flying on a *particular* flight – the passenger's first concern.

The accidents I have chosen for review here are intended to be as representative as possible, but the choice is not easy because, if the archives teach us anything, it is that there is no such thing as a typical accident. This lies at the heart of the argument that follows.

As we have seen, although the million-to-one risk of death in the air may be true as a world average, as a figure it has more validity for airlines and their insurers than it can have for the intending passenger who is committed to fly on a particular day, with a particular airline and a particular aircrew. The million-to-one odds may be shortening, as it were, even as he is driving to the airport. A number of other variables that can affect the odds then begin to take effect.

For a start, aviation insurers covertly admit that the premiums charged to different airlines vary considerably according to their safety records. So that, if we did but know how, the choice of airline is important.

A host of 'non-average' factors – unseen or not – that determine the safety of the coming flight will become evident from the accidents discussed in the chapters ahead. Amongst those that feature most often are the vagaries of the weather, the type and health of the aircraft, the physical and mental state of its aircrew and their ability to operate effectively as a team.

For example, a pilot who is suffering from the aftermath of a late night or who started the day with a domestic row over the breakfast table will not add to your prospects. He may fall asleep. If his co-pilot happens to be suffering from incipient dysentery following a foreign stop-over, the odds can shorten further. The pilots may have had a blistering shouting-match about union affairs in the crew room just before your flight. Along the route, a strike by air traffic controllers may lead to chaos and heighten the chances of a collision. These are some of the ingredients of risk taken from real life.

As the passenger fastens his safety belt before take-off, the notion of his million-to-one chances of survival is beginning to look rather theoretical and may have the smallest relevance to his particular flight. While world safety averages have their uses in the industry, they have proved to be profoundly attractive to the travel trade which has exploited them to such effect that many are led to believe that every ticket holder faces the same minimal risk. The record of events should invoke a wiser scepticism.

Towards a remedy

These remarks about the risk of flying have so far only dealt with the down-side: it stands to reason that if the dangers are greater on some flights, they must be less on others if the million-to-one world average is to balance out. Passengers are never likely to be told enough to choose the safest airline or the safest flight, but armed with a knowledge of the danger signals as revealed in the accident analyses that follow, one hopes that they may be better placed to ask the right questions. Informed inquiry is a powerful antidote to the blandishments of the public relations fraternity.

If this is a platitude, it leans on another that is none the less relevant. If companies owe their first duty to their shareholders, as they do, it follows that they have an equal duty to withhold any news of corporate affairs that is damaging to their reputations or financial standing, so far as the law allows. Yet the commercial ethos in which they operate purports a faith in the sovereignty of the consumer. The corollary that an ignorant consumer has precious little sovereignty in the market place understandably gets less attention. The orthodox answer to the competing claims of shareholders and customers is, I imagine, that the latter must inform themselves as best they can – in other words, they rely on the media to arm them with the facts. Thereby is sovereignty conferred. Apart from the limited disclosures required in company reports, other sources are scarce.

This may be no more than a glimpse of the obvious, but such an order of things bears heavily on the discussion of the role of the media in promoting air safety (Chapters 10, 11). However far the present wave of distrust of the media may be justified, it is argued that its role as a safety watchdog remains paramount.

Secrecy

It is explicit in the prime role of the media that information is hard to get. Secrecy, in all its forms, is considered in Part 2, although it inevitably spreads its tentacles throughout other sections of the book. The attention given to it in a study of air safety should, I hope, become evident from the context but a word of explanation may re-affirm its importance as a recurring symptom of accident.

The term secrecy is inadequate to describe the much wider collection of human motives that are at play: there seems to be no collective word that embraces all the elements of cover-ups, the constraints of commercial and bureaucratic confidentiality, pride and chauvinism, the fear of losing face or jobs, and the self-regarding urge for exclusivity often founded on misuse of the 'need-to-know' precept. All of these play a part among the accident cases considered here and this leads to the contention that – apart from honest and inadvertent mistakes – these covert motives prove to be the most potent source of accidents, and a prerequisite to most of them.

From this follows the call for greater freedom of information and a more vigilant media. There are many who oppose this view utterly.

The less that is said, they believe, the better for all of us. For example, a classic statement of this view appeared in a recent letter to the British magazine *Flight International* (28 January 1989), following the press coverage of two major air disasters. The writer complains that the lay media latched on to the incidents in order to weave lurid fantasies in the belief that this is what the public wants. He concludes that information about accidents should be regarded as *sub judice* until properly investigated by the competent authority: premature comment should be held in contempt, not only of the authority, but also of 'the respectable and respected professional standards within the industry'.

If there were no other reason for writing this book, the need to challenge this point of view from the record of accident causation alone would be reason enough. I suggest that those who adhere to the writer's beliefs in confidentiality should test them at least against the history of the Paris DC10 disaster and the first Boeing 747 jumbo to crash at Nairobi, backed by a sad number of other preventable accidents that are reviewed here. It is for them to judge whether or not a dose of 'premature comment' might not have saved many of the lives lost in them.

Terror

It should be noted here that there is one significant qualification to my argument about secrecy and that is the matter of terrorism. Over the past decade terrorism in the air shows itself as an increasing threat to the safety of civil aviation. That is the ominous fact seen from the record, sadly underwritten by the bomb that eliminated the Pan Am B747 Clipper and all 270 people within it at Lockerbie in Scotland on 21 December 1988 – Britain's worst aviation disaster. In that year, 578 was the highest-ever number of people who died in seven various forms of attacks on airliners worldwide, nearly doubling the total of 333 killed from those causes in the eight attacks during 1987. The average for the three years to 1988 shows that terrorism or deliberate attacks now account for nearly 40 per cent of all aviation fatalities worldwide (excluding the Soviet bloc and unknowns). Over a comparable period twenty years back, the record is almost blank: it shows but a single entry in March 1966 of an attempt to hijack a Soviet Ilyushin 18 at Havana in Cuba in which the pilot and one passenger was shot.

All that need or should be said on this subject is that the least that is said the better. From the bulk of information available to inquirers –some of it freely available and some of it highly confidential – there is hardly a fragment that is not of some potential interest to the intending terrorist.

There is some comfort to be drawn from the sure knowledge that both inside and outside the business of aviation, a very competent force of professionals is dedicated to reducing the threat. The surest method of hindering the terrorists is to ensure that they do so behind closed doors and that their strategy is kept secure. The temptation to relay some of the newest counter-measures is one that is willingly resisted.

Is fear justified?

If the premise is accepted that the true risk of an accident varies so much from flight to flight, can it be said that the apprehensions of so many passengers are irrational? The tendency to write off such fears as a known phobia is unconvincing because they arise from no single cause and so defy a slick diagnosis. Doctors say that the three factors involved are claustrophobia, vertigo (which can have physical rather than mental causes) and – something that defies a text-book label – a compulsive fear that the aeroplane will fall out of the sky. But this happens rather too often in reality for the passenger's forebodings to be described as an imaginary fear or phobia. Knowledge, it is argued, is the best antidote to unjustified fears.

Air passengers show a curious ambivalence towards the safety of their persons. There are those fatalists who don't want to know about danger at any price – the sort who put off going to their dentist or doctor in case they are told the worst. Arguably, this is a heroic stance, like the soldier who braves the bullets in the belief that only one with his name on it will find its mark. But they are outnumbered by the worriers – those whom the fearless fatalists might call the curious, the timorous or even the neurotic.

The hard core of the flight-shy is surprisingly large according to passenger surveys, which suggest that between 20 and 27 per cent admit to an abiding fear of the air. The industry has taken this to heart by offering thoughtful remedies to coax them aboard – British Airways, for instance, is currently running a £45 weekend Nervous Flyers Club at Manchester airport. More recently, two of the airline's captains and a London psychologist formed a company to offer a one-day cure at Heathrow airport for fear-smitten travellers that proved its therapy by ending with a flip in a sleek new Boeing 757.

These ventures should not lack for custom. A few months after they were set up, British tour operators reported a drop of 2.5 million holiday bookings in the aftermath of a spate of fatal crashes during the winter of 1988–9, most notably the Pan American jumbo destroyed by a terrorist bomb at Lockerbie, Scotland. The largest UK tour company, Thomson Holidays, said that there was no doubt that some passengers were becoming concerned about flying. One airline, which preferred to remain anonymous, reported many calls from customers who wanted to know the type of aircraft and details about maintenance procedures before they would book a flight.

Experience suggests that such phenomena tend to be short-lived as memories fade. Taken together, those passengers who suffer fear and those who feel none at all, are in the minority. It seems fair to assume that the average air traveller takes a healthy interest in flight safety and the secrets of the air.

Airlines clearly prefer their customers to belong to the fatalist camp. The last thing they like is any mention of the word safety, so that it is not surprising that the marketmen boost their chosen image of the mature and sanguine passenger who is bored out of his mind by such twaddle. This was affirmed by a discussion a few years back between the editors of two leading airline in-flight magazines and reported in *British Airports World* (a bi-annual produced by the British Airports Authority). The interviewing reporter quizzed them about the taboos that need to be avoided in airline reading matter –no sex, politics or religion. When it came to safety, the editors agreed that they well knew the 'areas of sensitivity' and avoided them. 'We don't want pieces on air crashes,' they said. 'Our aim is to entertain.'

As the buyer of last resort, the traveller finds that few commodities come under such inscrutable labelling as an airline ticket. The dim small print on the back page offers no pledges of safe carriage, but explains just how little you will get if you lose a limb, a wife or a

11

suitcase. Even if it were possible, no international consumer group has dared to list risky airlines or dangerous airports.

Picture the common-enough plight of someone trying to get home from some distant airport. Suppose that the weather is foggy, his flight has been switched to an unfamiliar airline, and air traffic controllers are on strike. At the bar, the impatient aircrew of the waiting plane can be heard having a row with the airport staff about the delay. A few days earlier a plane has crashed into the surrounding mountains soon after take-off . . . a not entirely apocryphal scene.

The ingénu who asks his travel firm about the safety of a flight is likely to be told that Madam and Sir have nothing to worry about because *all* airlines on their books are equally safe. We don't know whether the chairperson of the Axminster Ladies' Guild ever asked such a question when she stepped into the offices of Unicorn Travel in Bristol to arrange their day-trip to Switzerland. But if she had, the answer would not have foretold the fatal risks her party of shoppers were about to meet: the system took care to paper over the cracks.

Libraries are well stocked with authoritative studies on air safety but they are mostly written by professionals and for professionals. My credentials are rather different, as an aviation writer who has had a foot in many camps of the aviation business. If journalists have a claim to be heard, it is that they cannot help but become a clearing-house of ideas sourced from all levels in the industry. Since much of this input will be discrete data which is unlikely to be shared among other sources, the journalist's collected gleanings acquire a rare, if not unique, quality. He is rather in the position of a jury hearing testimony from a number of witnesses, none of whom knows what the others have told the court. The analogy may pitch the role of the reporter too highly, but whether he is regarded as a buzz-fly or a trash-can of trifles, his status as a professional bystander gives him a birds-eye view, if not so much as a panoptic vision, of affairs. At worst it may be shallow, but at least it is wide and, in sum, unseen to the other participants in the act.

On a personal note, I have to confess that my pedigree as an aviation journalist is impure. I came to the trade by default, as a lawyer in consumer affairs, to form the *Sunday Times* Insight Consumer Unit. Previous work had involved me in air passenger safety, and the editor of the paper soon detected that I was in the habit of flying aeroplanes. The transition to aviation writer soon followed, aided by editorial support of my aspirations as a pilot.

When my publishers asked me to write this book, I was encouraged to do so by the surprisingly large number of air-minded people who seemed avid to know more about flying. They showed a greater capacity to absorb the technical hows-and-whys than might be thought. This is not so surprising when it is realised that the English-speaking world is the leader in the game – America with the world's largest aviation industry, followed by Britain as number one in Europe. (The Eastern bloc is excluded due to lack of data, as it is throughout the book.)

There is a grim prospect ahead for those who resolve never to fly in an aeroplane. With close on 9,000 large passenger jets now in world service, it is predicted that 50 million new passengers a year will be needed to fill the 3,000 new jets now on order to the airlines. The pace of growth continues: the first passenger jet transport, Britain's de Havilland Comet, took to the air forty years ago. Only forty years before that flight, in 1909, Louis Blériot was the first person to fly the English Channel.

Omissions

A book such as this cannot pretend to offer more than a glimpse through the air safety scene, and some aspects have had to be omitted entirely. Others call for a brief mention.

The fleet of jet airliners that first came on the scene are now showing their age. Some are twenty years old or more and there is vociferous concern about their safety, despite the theory that every airliner is maintained to the standards of airworthiness now in force. A series of accidents caused by structural failures has shown too often that this ideal is not always met. I have not included the vexed question of geriatric jets because, by and large, it exhibits the same shortcomings in the policing of the airlines that appear in so many other contexts.

For example, the astonishing survival of the Aloha Airlines Boeing 737 that lost much of the roof of its cabin at 24,000 feet is a case in point. Only one person died, a flight attendant who was sucked out in the explosive decompression as the eighteen-foot section of roof ripped away. The aborted flight to Honolulu on 28 April 1988 made a safe emergency landing elsewhere. Metal fatigue in the 19-year-old aircraft was the cause. The US National Transportation Safety Board's report of the accident blamed the airline's maintenance programme. Contributing to the cause, it said, was the Federal Aviation Authority's failure to monitor the airline at its base: an FAA inspection of Aloha Airlines four months before the accident failed to inspect and report the physical condition of the planes, as it should have done. A similar pattern of events will be found to lie behind many of the accidents considered in the pages that follow.

There had been mounting concern in the industry about a series of fatal structural failures in ageing jets, but the dramatic pictures of Aloha's Boeing as it stood on the ground – looking almost as if it had been cut apart by a giant can opener – caught the public imagination with evident effect. Lufthansa, the West German flag-carrier, reported in September 1989 that it was receiving between twenty and fifty calls a day at its London office from passengers and travel agents asking about the age of its aircraft. With an average age of less than seven years, its fleet is younger than many of its competitors and the airline believes that this may explain a 25 per cent increase in its Atlantic business travel. It is seen as a customer-shift away from carriers using older airliners. Lufthansa quotes the travel trade's belief that media exposure of the dangers of old aircraft that became public knowledge as a result of the Aloha accident had made an 'enormous impact' on intending passengers.

While the dangers of ageing jets have declared themselves, at the other end of the scale modern technology has brought its own hidden perils. For example, however foolproof automated systems such as 'flight-by-computer' may be in themselves, their interaction with the human element has added new risks that were so clearly exemplified in the Mount Erebus tragedy in November 1979. All the 257 people on board the Air New Zealand DC10 were killed when it struck the summit of the mountain during a sight-seeing trip into Antarctica which had set out from Auckland, New Zealand. The official accident inquiry led to a political row and the appointment of a Royal Commission, although the complexity of the causes gives room for continuing controversy.

At first the pilots were blamed. The Royal Commission over-turned this finding. What can be called the navigational computer on the aircraft was routinely loaded with data for the Antarctic route by airline staff before the flight, but on the day of the accident the staff had altered the data, and so the route, without telling the aircrew what they had done. This, the Commission found, was 'the single effective cause of the crash' and the pilots were largely exonerated.

The approach

The arrangement of this book avoids the customary method of allotting chapters to each topic or theme, for reasons that govern my general approach throughout, and this calls for some explanation.

Accident analysis shows clearly that in nearly all cases causation appears as a chain of mishaps – defects, mistakes and a whole raft of variable factors – that culminate to lethal effect. A thematic approach that deals with each aspect under a single heading such as pilot error, engine reliability, weather and the like, can be an essential for the professionals and technical experts concerned with a particular subject. But for a general understanding of the whole train of events, it is fraught with absurdities and pitfalls. To pull out one particular factor from the context of each accident defies the nature of causation seen as a series of interacting events. This can too easily become an artifice bred of hindsight that, for my purposes, puts the cart before the horse.

A thematic treatment also comes close to a method that is open to a more serious challenge. I contend that it is manipulative, if not unethical, to adopt a theme and then set out to prove it by selecting snippets from the records in an adversarial manner. It implies a number of subjective choices down the line – the ranking of the importance of the chosen topic, what evidence to include or exclude, and how far to sever the material from its context. To do so introduces value judgments or prejudices that, like the clock that strikes thirteen, throws everything that follows into doubt. This method of selecting facts to prove a pre-ordained theme should be anathema to anyone engaged in the investigative process.

My approach is therefore rather to emulate the methods of the accident investigator who first considers the real-life imprint of events, and carries forward his findings to each successive accident in order to untangle the threads that are shown to be common to them all. Whether this objective or forensic synthesis succeeds or not is for the reader to judge. Where opinion or prejudice do intrude, I think

15

this should be self-evident. The method has its disadvantages. The result is not as tidy as the method of allotting a chapter to each topic, but I hope that the index and cross-references in the text will help to link specific issues where they arise.

The scheme of the book is therefore more like a jig-saw puzzle rather than a series of complete tableaux. As the pieces fall into place, the picture that emerges should outline the risks that face air travellers more nearly as they are, rather than as some would have us believe.

Part 1
The dangers of flying

1 | Pilot error

Pilot error still accounts for more accidents than any other cause, and this is likely to remain the case. Both in fact and figuratively, pilots are at the sharp end day in and day out, and by the logic of numbers they carry the highest risks and responsibilities. Few non-fliers appreciate the levels of skill and safety demanded of them, so that Dorman's almost self-inflicted lapse of airmanship at Basle is unusually shocking. Thankfully, his kind do not come often, but there are still plenty of examples of experienced airline captains who have been found to be unable to apply the basic principles of flying when they were faced with an unexpected emergency.

Basic principles

Leaving Dorman's infamous flight behind, let us take a leap forward to more recent times. Airliners are now far more sophisticated, of course. Complaints that pilots are becoming button-pressing systems managers rather than airmen, to the detriment of their flying abilities, seem to be borne out by the facts. Some pilots, in a lifetime of flying, may never have to face the dangers they learnt to cope with during their training. But the test did come for Captain Min-Yuan Ho, seated at the controls of his Boeing 747 jumbo as he prepared to land at Los Angeles on a wintry day in February 1985.

So far, the captain had piloted an uneventful flight across the Pacific from Taipei. At 55, he was one of China Airlines' senior pilots, with 15,000 hours' air experience under his belt, much of it spent on B747s. With about 40 minutes to run before the 276 people on board would land at LA, the aircraft was still cruising just above the tops of high cloud at 41,000 feet when it flew into some light turbulence. Routinely, Captain Ho switched on the 'fasten seatbelt' sign as the speed began to fluctuate uncomfortably in the rough air.

He called for reduced engine power, but the inner right-hand engine seemed to be misbehaving.

Leaving the plane on autopilot, he turned to discuss the problem with his other two aircrew and they decided to shut the engine down. When this is done, a plane will naturally slew to one side – in this case to the right – and begin to roll. However, while they pondered over the failed engine, he knew that the autopilot would correct the imbalance and keep the plane's wings level. It did so for a time. The system can cope with a rolling error of up to 22 degrees from 'wings level' and it compensated accordingly. But the three crew became so absorbed in their engine problem (which turned out to have been a false alarm: it need not have been closed down at all) that the plane's continued tendency to turn exceeded the autopilot's power to keep the wings level.

Captain Ho ordered a descent to 30,000 feet, the maximum height at which engines can be re-started. To do so, he disengaged the autopilot, still believing that the wings were level. The jumbo was already in a tightening roll to the right, but this was neither felt nor observed by the crew who were now descending through cloud. Their flight instruments could have told them the true attitude of the plane, but such is the power of auto-suggestion – unquestioned among the other distractions – that they failed to see the warnings before it was too late and Captain Ho began to lose control of the plane. Soon he was totally disorientated and, obsessed with the need to check the plane's speed as it plunged and turned, he made wrong corrective movements that only deepened his confusion.

Then followed a series of manoeuvres that no other B747 had ever had to endure. The huge plane rolled until its wings were vertical, then nosed downward at the start of a roaring descent. In its spiralling fall towards the Pacific below it turned completely upside down before falling into a vertical nosedive. These aerobatics created forces that not only severely overstressed the strength of the plane's structure, but at times left the crew unable to move as the centripetal forces pinned them down helplessly in their seats. As the jumbo gyrated downwards, the flight engineer fought to reach the controls but found himself pressed down in the aisle and unable to lift his left arm against a grip that at times reached 5G (five times the pull of gravity).

Normal rates of descent vary around 2,000 to 4,000 feet a minute, but the jumbo was falling at a rate of 18,000 feet (over 3 miles) a minute. By the sheer luck of the day, it broke out of cloud 11,000 feet above the sea. Captain Ho was now able to see the horizon and managed to level out the plane at a height of 9,000 feet. Amazingly, the battered plane had responded to the controls.

The flight diverted to San Francisco where it landed safely, with only two of the passengers seriously injured. Chunks of the plane's tail had been torn off, its wings bent permanently upwards and the auxiliary power unit (or APU – the small back-up engine mounted in the tail) lay loose and torn from its mountings. The cockpit voice-recorder that 'bugs' all conversations among the crew could have told much about their response to the drama, but it had been allowed to overrun and so wipe out the record so that we shall never know at first hand how it feels to fly aerobatics in a jumbo jet. But it is safe to say that it was a unique experience.

The official United States accident report adopts a suitably tactful tone in its handling of a mishap to a foreign state airline. It attributes the episode to the captain's failure to monitor the flight instruments which, it says laconically, resulted in his loss of control of the aeroplane. Such gentle euphemisms cannot conceal hair-raising convolutions that are more the sort of thing to be read about in the exploits of young pilots of World War One dicing with death over Flanders in their flimsy biplanes. How could it happen to a senior airline pro like Captain Ho, flying a load of passengers in the world's biggest jumbo jet?

The explanation needs some insight into the skills of blind or, more properly, instrument flying, the bedrock of airline training. Fun-fliers and novices in private light planes fly by day and in weather that allows them to keep the ground in sight – in other words, under visual flight rules. While a few private pilots qualify for all-weather flying, virtually all commercial flights must be conducted under instrument flight rules, day or night and whatever the weather. This is not only because the ground often cannot be seen from five miles up, nor even the horizon at night, but also because all such flights are guided primarily by constant radio contact with air traffic and radar controllers on the ground. Their job is to regulate the flow of traffic and give navigational assistance to pilots. It is this co-operative system that allows pilots to fly from A to B by using their flight instruments and without having to look outside the cockpit, except of course for the take-off and landing (and even these are becoming automated).

Such a system needs very highly trained piloting skills. Looking at the first obstacle that needs to be overcome, let us think back to the children's game of blind-man's buff. Victims are spun around until they reel and stagger; their sense of balance is telling them false. Balance relies on the three so-called semi-circular canals in the inner ear – one set in each plane, and all containing fluid. Bodily changes move this organ around like a spirit level and the upset is sensed and measured by hairs lining the canals. Too many rapid changes exhaust the 'memory' – the capacity to register and respond to them – and the system goes wild. Giddiness, or spacial disorientation, is the result: the brain sends out wrong commands to the muscles and other senses.

Flying 'blindfold' in rough weather puts your body through a whole chain of unexpected movements and so produces these symptoms beautifully. It takes a little longer in calm weather, but it happens just the same. Deprived of any visual references outside the aeroplane, the inner ear soon loses count and quits. But it doesn't tell you this, and the body sits there convinced it is upright when it and the plane may be plunging viciously off course or up or down. So in non-visual flight, the sense of balance is not only useless but worse than useless, 'because it becomes a compelling liar. Without instruments to put matters right, it is said that the average time for an untrained pilot to lose control and kill himself is between one and one-and-a-half minutes.

Recent research has shown that 15 per cent of military crashes and up to 35 per cent of non-airline civilian accidents (that is, smaller carriers, business and private flights) are caused by spacial disorientation. The survey does not deal with the airlines, but as we shall see, the same lapses of airmanship are prevalent enough at every level of the trade. Here is a clear example of how the classification of accident causes can be misleading: many fatal crashes tend to be written off under the heading of flight into bad weather because there is seldom any record of a pilot's final sensations. This research now supports the inference that the prime cause of many if not most of these accidents is not the weather but the lack of the instrument flying skills needed to fly through it. It is not easy to acquire them. It takes many sweaty-palmed hours of flight training to overcome these strangely compelling bodily feelings. On one occasion it may be that your senses shout at you that you are screaming down in a left-hand spiral, when the instruments insist that the plane is climbing madly to the right and about to stall. So pupils must learn to suppress nature's commands from the seat of the pants and obey the dials or perish.

Try for a moment to imagine being a pupil on your first vicarious instrument flight. Your training for this is done with forward vision and all but four basic instruments blanked off. Flying instructors need to be hard men. After screwing the plane through a quick series of giddy spirals that force your chin onto your chest and make your fingers feel like inflated bananas, the instructor will throw it into one of his choice 'unusual attitudes' with the command 'You have control!' Like hell you have – at least for the first few attempts. The nearest likely thing to this first sensation is a trip on a fast roller-coaster with your hands over your eyes. Sooner or later, by locking desperately onto the instruments, you may finally amaze yourself by emerging above ground the right way up.

How did the instruments help you to end up flying straight and level? When I tried to explain this in an article about an event rather similar to the China Airlines flight, it brought a lot of letters from inventive readers who suggested that a simple plumb-line suspended from the cockpit roof would give the pilot all the guidance he needed. One of these letters came from an old friend and ex-wartime Spitfire pilot – but I guess you are allowed to forget a lot in forty years of farming under the skies you once defended.

For those who don't often tangle with plumb-lines or resultant forces, an example may help. Imagine riding a bicycle in a long but tight turn, with a plumb-line tied to its crossbar. If the turn is kept steady enough, the string will point down to your feet, just as it would when you are riding upright in a straight line. In a perfectly balanced turn, centrifugal forces push the weight outward to remain aligned with the angle of the bike exactly as it is when riding straight, so it fails to indicate the turn. The plumb-line idea doesn't work.

The device that is needed to do the job is a top, or gyroscope. Once its wheel is set spinning, its position in relation to the ground will stay the same however much you tilt the base it may be resting on, and in whichever direction or plane you try to push it. A permanent line or bar fixed to its plane of rotation will remain parallel to the earth's horizon. And by the same token, it can also indicate any pitching in the forward and aft plane. Modern instruments are far more complex and go by many fancy names, but the principle of the artificial horizon holds true. It remains the centre of a group of vital instruments that make non-visual flight possible.

Captain Dorman had never mastered the art, or just enough to fool the examiners at his ninth attempt. Had Captain Ho forgotten his basic instrument training? The record shows that as an airline recruit, he had to be proficient in steep turns, recovery from the stall and low-speed manoeuvres. But 'recovery from unusual attitudes' is required neither by his airline nor, the accident report tells us, in the curricula of 'the major US carriers'.

This comes as a bit of a surprise to British and leading European pilots who are all drilled in such fundamentals. Maybe these other airlines take it for granted that basic instrument handling is built into every pilot from his early student days. If so, they are wrong many times over. Numbers of experienced pilots have crashed their planes because they were still capable of becoming disorientated. Thankfully, Captain Ho saved the lives of his passengers and crew. And he certainly did it in style.

But fate also took a kindly hand. Captain Ho and his crew had crossed much of the Pacific by night and through the dawn. It was broad daylight when, at 10.16 a.m., their drama began. Beneath them lay the mountainous clouds of a weather front with their tops at 37,000 feet and the base still at 11,000 feet. That morning's satellite weather pictures had shown conditions worsening towards the Californian coast, with a nasty tongue of lowering storm clouds moving out to sea. Had the accident happened a few minutes later, Captain Ho might never have emerged below the lower cloud-base over the ocean in time to level the plane out of its dive in the smaller layer of clear air that would have been left to him.

Given that he needed clear outside vision to regain control, the flight would have faced the same fate had it been night time. Only a month earlier, the 71 souls travelling aboard a night charter flight from Reno, Nevada, to Minneapolis on a gambling junket trip were not so lucky. As they took off into the night, they were not to know of flaws in both the plane and its crew. None of these weak spots was especially unusual or even potentially lethal, yet in combination they were soon to become so. Curiously, here too, the crew of the

Lockheed Electra had wrongly suspected engine trouble when the plane lifted off from Reno-Cannon International airport and the crew heard a 'thunk' as it became airborne.

'What is it, Mark?' the captain asked.

'I don't know, I don't know, Al,' responded the flight engineer. They were faced with the most hated crisis in a pilot's life – an unexplained fault in the first few seconds after take-off. Height and speed mean time, and time means safety. All these margins are at their very narrowest in the moments after the wheels leave the ground, in a phase of flight when even hard-boiled veterans mentally cross their fingers.

Both training and instinct call for a steady surge of engine power while the plane gains height and speed. If one engine should fail, optimum power is applied to the others to compensate for the loss of thrust. Short of an actual failure, airline drills may call for an engine to be shut down if it shows signs of a serious and developing fault. Again, the call is for more power from the good engines.

But less than half a minute after the four-engined Electra had taken off, the captain called for all throttles to be pulled back to reduced power. Suspecting that the 'thunk' might be an engine fault, he gave the fatal command: 'Okay, pull 'em back from maximum.' He turned to his co-pilot and told him to radio a request to turn back and land at the airport. In the darkness the Electra's altimeters showed its climb had flattened out only 250 feet above the ground, and a few seconds later the ground proximity warning alarm sounded in the cockpit. Meanwhile the co-pilot, who monitors the plane's airspeed indicator, saw that the pointer was falling back dangerously. 'A hundred knots!' he warned. The captain did nothing. Nor did his co-pilot. A few seconds later the co-pilot had to warn the captain again before the senior at last reacted and called for maximum engine power.

It was too late. Speed had decayed beyond the point of recovery and the plane was about to 'stall' – the moment at which a plane stops flying and drops to the ground (see pp. 27–9). A voice on the passenger intercom said tersely: 'We're going down!' Six seconds later the Electra crashed into a parking lot a mile and a half short of the airport. All but 16 people died in the impact, or were too badly injured to move. The fire from 2,000 gallons of fuel on board killed 15 of those who had survived the impact of the crash.

The sole survivor was a 17-year-old boy who had been sitting next to his father. He was thrown clear and landed 40 feet away from the blazing inferno, still strapped to his seat. Although badly hurt, he ripped himself free and ran for his life. Later he told rescuers that he had been thrown through a bulkhead – 'I'd seen movies on how to prepare for a crash, but I didn't think bending over in a prone position would help, so just before impact I covered my face with my arms and pulled up my legs.'

It was a short-lived but horrible tragedy. Many may question the need to revive the memory of it or ask what possible justification there can be for raking over the embers months after the event. It is a plea that comes with added force from those bereaved by fatal accidents, and it demands a sensitive answer.

The facts that came to light long *after* the Reno embers had cooled provide the only possible justification, just as in the case of those that were revealed after the inquiry into the crash in the Swiss mountains. We should feel safer for that knowledge.

The engines of the Lockheed Electra can be started by an external air-pressure line supplied by airport ground crew. This is inserted through an access door on the wing-top between the cabin and the inboard right-hand engine. On the night of the Reno flight, that door had not been properly fastened after the start-up and it had blown open as the plane gathered speed.

After the crash, it came to light that this had happened before to other crews, who had reported that the resulting vibrations caused by the door buffeting in the airstream did resemble an engine fault. But the Reno crew never had the benefit of that experience. Had they known or been told about it, their false – and fatal – diagnosis of engine trouble could have been avoided. Ironically, the opening of the access door is not critical and the Electra could have been controlled safely with it open in flight. Here too, the crew had not been told.

The opening of that access door eventually released a great deal more about the history of the flight than the part it had played in triggering a chain of errors by the Electra crew. Had there been no cause for a post-mortem into that flight, it might never have been known that the captain was angry because the flight had been held up for three hours; that the crew were so 'careless' in carrying out their safety checklists that ten items were skipped, others done in reverse order, and fourteen more not called out at all; and that the captain omitted altogether his mandatory pre-flight briefing to his crew.

The accident investigators found that the captain's failure to follow these and other safety procedures could have happened because the flight crew allowed themselves to hurry. They criticized other conduct which, although it may not have influenced the outcome of the flight, showed astonishing laxity: the crew neither bothered to check the loading and balance of the plane, nor did they compute the vital weight-and-balance safety limits before take-off – a compulsory and life-preserving necessity before every flight.

The final verdict of the investigators declared that the captain's action in reducing engine power and the whole crew's failure to monitor the plane's airspeed, as well as other primary lapses of airmanship, all led directly to the accident. In other words, in the heat of the moment the crew simply forgot to watch the most vital danger signal of all – the needle of the airspeed indicator falling back beyond the point at which the plane could remain airborne. The event of a 'stall' such as this is too common and fatal an occurrence to be described in a sentence. To know your lowest safe airspeed could be said to be the first commandment of every aviator for fairly obvious reasons. But the reasons why a stall spells such sudden disaster calls for an explanation.

Stalling

In this context of course, the word has no connection with the 'stalling' of a car engine. Here it relates to the flow of air over the wings and body of aircraft. There is a neat way to show the seemingly perverse effects of a fluid flowing over a curved surface or an aerofoil shape such as wing.

Suspend a large spoon lightly by the tip of its handle near a running tap and move the hollow of the spoon towards the column of water. As you would expect, the water pushes the spoon away. Now turn the spoon around to bring the convex curve or bulge to touch the water and it may surprise you to see that the spoon will be sucked *inwards* towards the flow.

So when a flow of fluid is diverted over a convex surface, it sucks the surface of the object inwards towards the disrupted flow. It follows that a horizontal flow of air over the top of a curved wing will suck it upwards, thus producing most of the 'lift' needed to take a plane off the ground (only about a third of this force comes from the push of the air under the inclined surface of the bottom of the wing).

Put simply, that part of a moving column of air forced to flow over the longer route of a curved surface, has to speed up before it rejoins its parent column beyond and behind the obstruction. The accelerated flow that sucks the wing upward must be fast enough to keep a smooth unbroken film of air sliding over its surface. If the airflow slows up, there comes a sudden moment when it ceases to adhere and breaks up into a confusion of eddies. The sucking effect or lift ceases as abruptly, while the broken flow produces added 'drag' with its horizontal braking effect. The plane then becomes a descending object rather than a flying machine.

The suddenness of a stall can be quite dramatic in some high-performance and training aircraft – a quick snap and the nose plunges down. (One memorable type, clearly designed to teach through terror, stalls with a whack that fills the cockpit with a blinding dust to add to the trainee pilot's ordeal.) Unless the stall is checked, its sequel will be an uncontrolled spin into the ground.

Modern aerodynamics have made the stall more docile, usually preceded by a buffeting of the airframe that warns of its onset. Yet, as will be seen, it still happens.

The exact stalling speed on a particular flight varies according to many factors such as the load carried and the configuration and manoeuvres of the aircraft. Extending the wing-flaps at take-off and landing, for instance, reduces the effective stalling speed.

The smooth flow of air over the wings depends not only on speed but also its direction. Normally the airflow is horizontal. If it meets the aircraft from any lower angle, it will have to curl round further before flowing over the top of the wing. This disruption causes the break-up of the airflow over the top of the wing to occur sooner, so that the point of the stall depends on the angle between the path of the aircraft and the direction of the airflow that is meeting it: this relative angle is known as the 'angle of attack'.

For example, imagine an aircraft flying in level cruise when its power is reduced. If its pitch attitude remains unaltered, it will begin to sink. The relative airflow is now coming from underneath as well as ahead so that the angle of attack is increasing towards to point of stall. The pilot can avoid this either by pitching the nose down, or

28

increasing speed to raise the force of the airflow over the wings. In other words, as the plane descends his effective stalling speed increases.

Conversely, if a pilot flying level at high speed throws his plane into a violent climb, the high angle of attack can catch him out and cause a rapid stall: the relative airflow meets the underside of the wings and can no longer flow smoothly over their tops. In this extreme case, his effective stalling speed has leapt up to meet his actual airspeed and a 'high-speed' stall' ensues.

Even in airline flying the risk is ever-present for the unwary. Changing winds and turbulence during landing and take-off can add to the danger. But as on the flight of the Electra at Reno, the over-riding cause is human oversight and this is usually induced by other distractions and stresses on the flight deck.

Perhaps a pattern is already beginning to emerge from the three accidents so far considered here: human failures, but set in a context of apparently random kinks and flaws beyond their control. Breakdowns in the performance of aircrews can be better understood by looking briefly at their complex theatre of activity, the flight deck.

The flight deck

It may seem strange that an entire crew of three or four airmen can fail to spot the source of the main danger that is overtaking them unawares, but it is a well documented syndrome. Stress can breed a fixation on one problem to the exclusion of others, and with it the loss of a sense of priorities. It is a human trait that shows itself clearly during instrument or 'blind-flying' training.

A central group of primary flight instruments has to be scanned at short but regular intervals and the flow of visual data from them translated by the pilot into frequent corrective movements on the controls. When these demands begin to exceed a pupil's ability, he tends to develop a psychological 'set' (see p. 172) or fix on one or two evident aberrations at the expense of ignoring others that are equally vital. He may, for example, try to check an incipient climb above his assigned altitude, and then ignore the rise in speed that accompanies his correction; or he may wander off track by allowing a slight banking turn, forget to tune and check a navigational instrument, or miss his cue for a routine position report.

He is therefore taught to maintain and comprehend the full scan of instruments until it becomes second nature. Only then will he acquire the spare mental capacity needed for other duties and excitements on the flight deck – radio chatter, following charts and checking flight logs. Such skills are akin to the agility of the juggler who keeps several balls in the air while delivering his patter to the audience.

When two or more are gathered together, the obsession of one seems to become infectious. Crews are of course taught to monitor each other but – sometimes even when things are going quite smoothly – a doubt or query from one draws the attention of the others into a kind of mutual tunnel vision. Possibly this underscores the saying that a trouble shared is a trouble doubled.

This appears to have been the case on the night of the British Midland B737 crash onto the M1 motorway in Leicestershire, England, in February 1989 with the loss of 47 lives. (See, more fully, pp. 171–4.) It followed a breakdown in one of its two engines. The final report of the accident is likely to confirm the conclusion that the crew inadvertently shut down the right-hand engine, which was sound, instead of the faulty left engine. As a result, when they tried to apply power as they came in to land, none was available from either engine and they crashed short of the airport. The puzzle is that all their instruments showed the fault to lie in the left engine, but for reasons yet to be fully understood, the two-man crew convinced each other that the problem lay in the right engine.

Potomac

Most of us will find it hard to forget the ghastly scenes replayed on our television screens of victims struggling for life in the ice-packed Potomac river at Washington in January 1982. Only 5 of the 79 souls on board the Air Florida Boeing 737 survived that winter ordeal. Their plane had failed to climb away from the airport through a blizzard, mushing down onto a bridge less than a mile from the runway before sliding under the frozen river. There had been the usual build-up of adverse factors and the subsequent inquiry into these taught much to prevent a similar recurrence, but there was one human response that still reads across the years to us today.

A heavy snow-fall had caused a build-up of traffic at Washington

National Airport that night and the Air Florida flight had been held on the ground for 45 minutes in the queue for take-off, while ice and snow gathered on its wings. Slush was building up on the runway when at last the crew were told to make an immediate take-off to clear the runway for the next incoming plane. By now the crew were fretting about the weather and the likely extent of the snow and ice on the unseen wings. That did not help them in their coming predicament, but it was not the determining factor. All the ingredients of stress were present on the flight deck.

As they accelerated down the runway, their flight-deck engine instruments told them that full take-off power was being delivered, when in fact this was not so. Each engine was programmed to deliver 14,500 pounds of thrust, while in fact they were only producing 10,750. Power sensing probes in the engines had iced up, and the known result of this is that the instruments over-read to show more power than is actually being delivered by the engines. Heating elements around the probes prevent them becoming iced up, but the crew had not switched them on before take-off, as they should have done. The accident inquiry found that this was a direct cause of the accident. But we have not yet reached the factor that could still have saved the flight.

Take-off power is pre-set and measured by one set of instruments (one for each of the two engines), the engine pressure ratio or EPR gauges. But grouped with these are four other sets of instruments that reveal the behaviour of each engine. A proper scan of *all* these indications would have revealed that the engines were not delivering and immediately needed more power: all the crew had to do to climb away safely was to push the power levers forward. The inquiry found no positive evidence that the captain had checked all his instruments and presumably never spotted the cause – the lack of sufficient power. Had he done so, he would have slammed the power levers forward, but they stayed at their faulty low setting to the end. The US Safety Board's report comments that 'the crew's action may have been due to a lack of winter operating experience, a lack of understanding of turbine engine operating principles, and perhaps, deficiencies in their training regarding winter operations'.

Certainly this follow-my-leader syndrome seemed to afflict the crew of the Eastern Airlines TriStar that flew into the Everglades swamp eighteen miles beyond Miami airport in December 1972. Its 'black box', recovered from the mud later, told of another needless tragedy caused by a relatively minor distraction on the flight deck.

When the undercarriage was lowered during the approach to Miami, the green warning light that confirms that the nose wheel is properly extended and locked down failed to illuminate. Miami control allowed the crew to break off its approach and bypass the airport to the west at a height of 2,000 feet while the crew sorted out their problem. From inside, all they could guess was that either the nose wheel had not locked down safely (which in fact it had), or that the warning light was faulty (which it was).

This drew the three men into an animated discussion, while the co-pilot searched for a pencil to prod the suspect lamp socket. Then he dropped his pencil and called for a torch to find it on the floor. The other two obliged and directed his efforts as he grovelled about under the instrument panel.

All the while the TriStar, left to its own devices, was continuing on its original auto-descent command to land – it had not been cancelled at 2,000 feet as it should have been. The captain looked out at the last moment to see the ground rushing up at him and he slammed on full power. It was too late to make a safe overshoot and they flopped into the swamps beyond, killing 99 of the 176 on board.

A less celebrated but more prolonged example of a team that succumbed to distractions ended in another avoidable disaster six years after Everglades. This time it struck the crew of three flying a four-engined DC-8 into Portland, Oregon, in the Christmas week of 1978.

Again the crew had suspected a fault in the landing gear as they made their approach to the airport, but as they had ample fuel reserves they decided to circle and check out the system. This they did for no less than a full hour before the flight engineer (FE) became edgy about the amount of fuel remaining, but Captain Malburn McBroom was unperturbed. The last pages of the transcript from the cockpit voice recorder spell out how wrong he had been. Some of it, in résumé, included the following:

[Eight minutes before the crash.]
FE: 'You've got another 2 or 3 minutes [of fuel].'
Capt: 'Okay.'
There is more talk about the mischiefs of the landing gear, then the
FE repeats his fuel warning. 'Okay,' says the captain, and resorts to
the gear again.
[Seven minutes to run.]
Co-pilot (CP):   'I think you just lost number four [engine],
                        buddy . . .'

Some idea of the rising tension is felt from other assorted remarks
around the cabin as time and fuel run low:

                        'Better get some [fuel] crossfeeds open there or
                        something . . . all righty . . . we're goin' to lose
                        an engine, buddy . . . why? . . . we're losing an
                        engine . . . why? . . . fuel . . . open them
                        crossfeeds, man . . . showing fumes . . . it's
                        flamed out . . . we're going to lose number three
                        [engine] in a minute . . .' Then there is an
                        argument about why the fuel is so low.
Capt: 'Get some (!—) fuel in there, you gotta keep 'em
                        running, Frostie.'
CP: 'Get this (!—) on the ground.'
Now with four minutes to run, they lock on to the instrument
landing system beams. For a minute they talk about the suspect
landing gear. Three minutes to go, then:
FE: 'Boy, that fuel sure went to hell all of a sudden.'
Capt: 'There's ah, kind of an interstate high . . . way
                        type thing along that bank on the river in case
                        we're short [of the airport].'
The stream of words become staccato:
                        'Let's take the shortest route to the airport . . .
                        about three minutes? . . . yeah . . . we've lost
                        two engines, guys . . . Sir? . . . we just lost two
                        engines, one and two [both left wing engines]
                        . . . you got all the pumps on and everything?
                        . . . Yep . . . *they're all going* . . . we can't make
                        Troutdale . . . we can't make anything . . .'

33

[One minute to go.]

Captain McBroom: 'Okay, declare a mayday.'

The last words on the tape come from the co-pilot's radio call to the airport:

> 'Portland Tower, United 173 Heavy [big aircraft], Mayday . . . we're . . . the engines are flaming out, we're going down, we're not going to be able to make the airport.'

Twenty-five seconds later the sound of impact cuts off as the tape stops. The DC-8 with 189 people inside it had crashed into houses among Portland's suburbs six miles short of the runway. Luckily there was no fire and only 10 people died, but 23 were severely injured. Both pilots were among the survivors. The flight engineer died.

Does pilot error cause most accidents?

It is generally accepted that pilot error causes more accidents than anything else, but there are some recent statistics that seem to challenge that conclusion. Certainly pilot error is often an imperfect statement and it can be unfair and inaccurate without some qualification. While it is generally agreed that the broader term 'human error' – which of course includes mistakes by pilots as well as air traffic controllers, maintenance staff and the like – ranks as a first cause, there are reasons to suppose that the part played by pilots has been quite seriously underestimated. It comes about from the way that statistics are translated into language.

For this reason there is no need here to delve far into the numerical content of statistical tables, but rather to look at the pitfalls of turning numbers into words. Purely to illustrate the point, take a set of figures from a recent US NTSB (National Transportation Safety Board) survey that lists accidents under the heading 'broad cause/factor' by percentage order of magnitude. Pilots score 54. 'Personnel' (other than pilots), 46, and bad weather, 42. Then follow ten other smaller categories including hardwear faults in landing gear, engines, airframe, instruments and equipment. The total adds up to more than 100 per cent because more than one factor may be judged a 'broad cause/factor' in each accident and so overlap.

The first simple point is that, at first glance, the list appears to put pilot error in first place. But if other causes, such as all mechanical failures, are added together they exceed the score for pilots. So you might think the machine more fallible than the man. Yet any accident may be judged to include more than one broad cause – for example, an engine failure followed by the pilot's mishandling of that fairly common event. So the same accident has contributed to more than one category of cause. Many other such tables listed under headings (or 'fields' in computer jargon) would pose the same questions.

It is that human value judgments conceal themselves in the way that the 'broad cause/factors' are assigned, as well as in the conception of the categories into which the figures are grouped. Look, for example, at the Vieques Air Link crash off Puerto Rico (pp. 57–61), and identify the originating cause: was it water in the fuel, the failed engine, or an incompetent pilot? Each of these could be legitimately slotted under the appropriate statistical heading and so fail to provide a clear answer.

The recognition follows that there is a difference in *kind* between the human non-statistical activity of measuring the significance of an event – and so determining its claim to a particular category – and the creation of purely statistical data such as, for example, the number of daily airplane movements at Chicago O'Hare airport.

The result seems to be that once a dose of subjective choice is injected into what appears to be innocent statistics, even worse complexities arise at the next level when 'factual' comparisons are attempted. It appears, from the first NTSB example above, that pilots head the list. Then they don't. And who decides that a particular accident classed as a landing-gear fault was not caused by sloppy maintenance and so should or should not swell the total of 'personnel' failures?

The same logical or linguistic false quantity can be seen from another angle. The question 'Does pilot error cause most accidents?' is answerable in statistical terms because it hides an assumption that air crashes have a single prime cause, while history shows that there is usually a miscellany of causes. And any attempt to quantify the relative significance of each contributory event is a subjective and not a statistical process. In other words, once the purity of valid statistics has been contaminated by a non-factual input, everything downstream of it becomes muddy and obscure.

The kind of question above about pilot error – one that is so often asked in one form or another – can therefore only be answered in its own terms, and those can only be subjective. So that it seems to me fair to offer an educated guess. If I were mentally to thumb through the 200 or so major accidents studied over recent years, my conclusion must be that pilot error appears to be the most common cause, and this is greatly fortified if those accidents in which it makes a sizeable contribution are included. A more authoritative conclusion is provided by the Lufthansa World Accident Survey in Appendix II.

It is true that mechanical failure added to or preceded by faulty maintenance may now be jostling pilot error out of first place. That indeed was the conclusion given to the Flight Safety Foundation in October 1986. The initial source in most cases, it said, was faulty maintenance – 'If something hadn't gone wrong with the aircraft in the first place, no accident would have occurred.' This leads us straight back into the deep waters of causal theory and invites the proper question 'but in how many cases should the flight crew have been able to remedy or overcome the defect and then failed to do so?' That is what pilots are trained to do . . .

Lastly, there is the rather glib category of weather as a cause of accidents. It is hard to find more than a few weather-related crashes in which the pilots do not share most of the blame for flying into it in the first place. With few exceptions it is eminently avoidable, but people prefer to take chances to get home to base and meet schedules. There is revived concern at the number of crashes caused by landing and taking off in thunderstorms and bad weather. The trend is sharpened by the recognition that most such accidents need never have happened – the Washington snowstorm of 1982 that led to the ghastly crash of the Air Florida Boeing 737 in the frozen Potomac is a classic instance (see pp. 30–1). Bad weather is a predictable risk that can be met by patience, a diversion or delay. But the statistics can, if misread, seem to tell us that it is the sole culprit. Picking on even the most authoritative comment out of context can be equally misleading.

In May 1987 a headline told that NTSB chairman, Jim Burnett, had said that bad weather near airports 'is *the* major threat to aviation today'. He went to show that in five years it caused 64 per cent of major crashes with nearly 300 deaths. That appears to put weather and not pilots at the top of the list, until you go on to read that his final advice to pilots is to *avoid* hazardous weather.

If it is judged that pilot error is indeed a prime cause, it needs to be re-stated that this does not mean that as a class flight crews are not performing close to the highest standards that can be expected of mortals. Far from it. From their leading role in the drama, the wonder is that they show such a remarkably low degree of fallibility. There are always exceptions. The treatment of pilot error and weather as causal factors exemplifies the dangers of turning figures into words. As it has been said, the interpretation of accident statistics as a whole is suspect for the same reasons.

2 | Insights into the cockpit

Studies of air safety are open to the challenge that they necessarily focus on the relatively tiny number of trips that come to grief and ignore the thousands of worldwide daily flights carrying their passengers in perfect safety. The subject must indeed be kept in that context, however obvious it may be. A kindred doubt is that the concentration on accident flights alone can breed an obsessive tip-of-the-iceberg attitude that uses only the worst-case examples as a measure of what goes on in the industry across the board. To do so, it could be said, is to treat the mishaps and errors revealed in the texts of accident reports as typical rather than atypical. In fact though, runs one argument for the defence, if every flight were to be subjected to the same detailed scrutiny, far from revealing a catalogue of blunders it would demonstrate the extremely high levels of safety that are being achieved unseen on the vast number of non-accident flights.

It is a persuasive view until it is met by the widely accepted proposition that most accidents are caused by a chain of mishaps or defects: it is when these reach a critical combination that an accident occurs. From this it would follow that an unknown number of flights carry some defects but less than the critical amount – a putative graph might show the fewest flights with a high number of faults rising steeply to the many with only one or two. But that could be no more than an academic exercise because the statistics are wanting and nobody really knows what may be happening on all those 'safe' and accident-free flights.

Fatigue

Here we are forced back on to anecdotal evidence which, however credible, can only provide a glimpse into the unknown. Take the recent case of an apparently normal long-distance flight that landed its passengers safely at Heathrow. Half-way through the trip one of the pilots had fallen asleep for 30 minutes, and when he awoke he said that he 'realized I was the ONLY ONE awake' on the flight deck. That crew had all been on duty for more than 12 hours. This would be worrying enough if it happened to be an isolated case, but it is clear than sleepiness among aircrew is far more common than official records might suggest. Instances are unlikely to be reported there for obvious reasons, but other sources tell a different story.

If it were not for its serious overtones, the words of one co-pilot on a flight from Gatwick to Africa some three years ago could read as a tragicomedy that luckily turns into a happy ending. His jet's departure was scheduled for early afternoon but the engineers had found a snag which, they said, would be fixed in a hour. It wasn't, and the crew kicked their heels for the rest of the day before finally making a midnight take-off with the co-pilot at the controls. Before the autopilot can be engaged, another auto-system – the yaw damper – must be selected. This prevents the tail of the aircraft from developing a sideways oscillation, a dangerous tendency that occurs in high-altitude flight. The system automatically deflects the rudder to counter this effect. When the co-pilot switched it on a few minutes after take-off, the damper misbehaved and gave him a crazy full deflection of the rudder. After switching it off, he radioed his company, who asked him to please continue the flight – 'We'll fix it on return tomorrow.' This led to a more tiring trip ahead, because the plane had to be flown manually all the way without the help of the autopilot.

After a refuelling stop in North Africa, the co-pilot found himself flying over the star-lit Sahara in the relaxed atmosphere of the warm dark cockpit. Laconically, he then describes how he fell asleep with a hundred tons of aircraft travelling at over 500 m.p.h. in his hands. 'The captain awoke first, punched me in the shoulder, scared the hell out of me. We all awoke, very sheepish . . .' The crew made themselves soup as they compared notes about their disturbed sleep-patterns before the trip. Still at the controls, this seems to have induced the co-pilot to fall asleep again – 'I awoke with a start to find the captain and engineer both asleep – I got my own back by punching my captain's shoulder.' Severely shaken by their behaviour, the three crewmen managed to keep themselves awake for the rest of the flight by switching on the cockpit floodlights, and they finally landed safely at their destination.

More recently, another pilot tells that twice in one year (1985) he had a captain fall asleep on a day flight. He goes on to say that in the last two years tired pilots have caused a noticeable increase in small mistakes such as forgotten safety check-lists and failures to level out at the altitude assigned to them by air traffic control. He means, of course, that such slips may be few but each of them carries the potential of an accident. He cites as another example the last leg of a tiring flight from an Italian airport where he failed to check the fuel uplift, with the result that the plane took off with an unknown extra load of $7\frac{1}{2}$ tons. He lived to tell the tale, but with a more fully laden aircraft and under less forgiving operating conditions, that could have been a fatal mistake.

Another close call came to two drowsy TriStar L-1011 pilots who had not had a full night's rest for four consecutive nights, and allowed the three-engined jumbo to come within an inch of stalling as it began its approach to an airport: although the autopilot was controlling the throttles, the captain had overridden them and pulled them back to the idle position. This happened while his attention was diverted to the co-pilot, who had not flown into that airport before and was searching for a sight of it. The speed fell to a dangerously low 133 knots (153 mph) before another of the crew noticed it and shouted 'Speed!' – prompting the captain to slam the throttles fully open just in time to keep the plane from stalling.

All of these incidents and others echo a growing number of complaints about fatigue on the flight deck: 'Nearly all the captains I fly with are noticeably tired and fed up and there is a general air of lethargy on the flight deck. . . . All three crew members were finding it very difficult to stop nodding off. . . . Something MUST be done soon to change this before there is an accident . . .' The good news is that these warnings are now being voiced. The question is how far they are being heeded.

There is a perennial debate about fatigue on the flight deck between the pilots and the authorities who lay down strict flight-time limitations on all airlines. Since none of these incidents in fact broke those limits, the argument that has flared up once again is about where the line should be drawn between acceptable fatigue – as a normal exigency of the service – and the degree of exhaustion that begins to affect performance and safety. The pilots' side of the case, as exemplified by these complaints from the flight deck, come from a confidential 'confessional' run by the Royal Air Force's Institute of Aviation Medicine in Britain. Somewhat characteristically, this is modelled on a similar scheme that had been running successfully for some years in the United States under the auspices of NASA (the US space agency) and known as the Aviation Safety Reporting System (ASRS).

CHIRP (Confidential Human Factors Incident Reports)

The British project, now five years old, has invited pilots to contribute all their fears and errors in a strictly anonymous form and these are edited and then circulated to other airline pilots. Readership is high and as one pilot remarks, it is one of the few aviation circulars 'to be read from cover to cover by most pilots'. The benefits of learning from the mistakes of others are obvious (there, but for the grace of God, go I) and it is hard to imagine any other initiative that could improve airline safety so simply and directly. The only wonder is why it came so late in Britain. The CAA, seeking an independent group to inspire the confidence of pilots, farmed it out to the Air Force's medical research team, who have adopted a suitably chatty and informal style among the pilot ranks (it invites 'personal cock-ups' that might be too embarrassing to report elsewhere). The pages of CHIRP, as it is known (Confidential Human Factors Incident Reports), do of course contain other pilot worries besides the fight against sleep, but fatigue appears as the most common source of error on the flight deck.

Only a fraction of the large response from pilots can be circulated. An editorial summary relates that reports have come in from all sectors of civil flying about whole crews falling asleep on the job, while others reveal the link between piloting errors and the low motivation brought on by tiredness. More disturbingly, the circular reports alarm at the number of telephone calls from crews who 'sound at their wits' end with fatigue and frustration because they feel that nothing is being done for them, yet feel unable to tackle their companies for fear of being branded as troublemakers and jeopardizing their jobs.' Many pilots realize, comments CHIRP, that their companies are on a hiding to nothing because they feel compelled by commercial pressures to operate to the legal limits 'and they can't be held responsible if the limits are too liberal'.

41

CHIRP has now won acclaim and notice in high quarters. In a recent issue the CAA's director general of operations comes to the defence of its flight-time limitations as they are now set by law. He admits concern at the complaints about fatigue – which, he explains, are under constant study at home and abroad – but goes on to point out that random factors can disturb pilots' rest periods. Such individual cases, he says, cannot be dealt with by changing the general legal limits. The DG then reaches a broad conclusion that strikes a note of truth that should be heard in any discussion about air safety. 'There is,' he writes, 'no field of human activity immune from the occasion when, once in a while, *all the adverse factors pile up together* to create a situation no one would want to see' (my italics). He concludes that no set of rules can eliminate these cases completely.

Taken in a wider context, these words are an apt description of what is probably the most common accident scenario. They reinforce the multiple-cause theory which in turn suggests a wide spread of 'adverse factors', in varying numbers, which are present but unseen on many flights and only reach a lethal total on the very few. While this contrasts sharply with the orthodox claims heard in the industry that every flight is as safe as can be, bar one in a million, it supports those safety trouble-shooters who know that there are likely to be some potentially lethal factors present on many, if not most, flights.

If this sounds alarmist, it is at least a healthier perspective to bring to each and every flight. The job of detecting and eliminating the danger factors – even down to the routine read-out of pilots' checklists – becomes a positive rather than a negative activity when faults are expected. From a host of other similar instances that could be cited, there is the jumbo that crashed at Nairobi in 1974 because the leading edge flaps were not extended before take-off (pp. 118–27). The pneumatic valves to power them had not been turned on because the crew muffed their checklist – the recorded voice of the co-pilot is heard to call out the item rather indistinctly, but there is no positive reply and there should have been.

After fatigue on the flight deck, the other most frequently reported topics listed by CHIRP are air traffic-related incidents (ground controllers have now been invited into the confessional) and simple pilot error of the kind shown in the Nairobi take-off sequence. An eloquent instance comes in another chilling account given by the flight engineer on board a four-jet passenger flight that ended in a dizzy landing which he saved from disaster by seizing the initiative from the pilots. His elderly captain had been told to land on the left of two parallel runways at a busy international airport, but as he passed over a position fix some four miles out on the start of his final approach to land, the plane was flying too high and too fast. Warned of this by radar controllers, he tried to increase his rate of descent as the crew reminded him that no safety checks had been done, nor were the flaps and undercarriage yet lowered for landing.

At this point the airport controllers asked the captain to land not on the left but on the right-hand runway. This threw him into greater confusion and he slewed the plane into a high angle of bank to the right, so that when it crossed the threshold of the runway – at 300 feet and much too high – it was pointing at an angle of 45 degrees to the runway direction and heading towards the terminal buildings. Added to this, in his attempts to get the plane down, its vertical rate of descent as it came to meet the ground was far too high. His approach, now wildly beyond any safety limits, was heading for certain disaster and screamed for an immediate overshoot – the flight engineer had already called for this three times, but to no avail. Meanwhile, he noted later, the co-pilot – the third member of crew – had not uttered a word and seemed content to let his captain steer them into a crash landing.

At the last minute, the engineer forced the issue by pushing the throttles forward and calling for an overshoot – and the captain, apparently shaken into reality at last, accepted and repeated his command. The jet climbed safely away and as it circled for another attempt, the captain thanked his engineer for making the right decision. They eventually landed safely at the second attempt, but the engineer records some unresolved doubts.

Why, he asks, was the co-pilot so overawed by the captain that he would have died before questioning his disastrous approach? And why did the captain wait for someone else to act when he must have known that he was heading for a crash landing? It has to be remembered, of course, that there are two unheard witnesses from the flight deck who may have answers to these questions. The switching of the runways at a critical stage of the approach by the airport controllers is an unwelcome practice at any time, and more so when they knew that the flight was already in difficulties. But the captain's generous acknowledgment of his engineer's saving action

43

adds credence, if it is needed, to one man's memory of a hair-raising episode.

Just at what point a junior should override a senior member of crew is a constantly recurring theme in flying lore. It always takes guts to criticize the boss, even when the chips are down. The theory is perfectly clear: when in doubt – shout. But in practice, the fear of a wrong call or a resentful or domineering senior can inhibit that ideal, as too many accident reports testify.

Recent events, however, have cast an ominous shadow not only over the survival of CHIRP as a voluntary confessional, but also over the compulsory reporting of airmisses by pilots and controllers. As a generality, many foreign jurisdictions prosecute individuals under the criminal law for negligent acts or mistakes that cause death or risk life – faults that under Anglo-American law would not amount to criminal offences unless they are shown to be either deliberate or grossly negligent. Recall the fate of Captain Arenas, for example, who in 1973 was instantly arrested after the crash at Nantes in France on a false assumption of guilt (pp. 91–5). Peremptory arrest has always been a threat to pilots involved in accidents abroad, but there has been an increasing willingness by European transport authorities to prosecute anyone who makes a mistake, whether or not lives were lost.

In May 1989 an Icelandic court heard criminal charges against those held responsible for a near-collision that had occurred three years earlier between a British Airways Boeing 747 and a Scandinavian Airlines DC-8. In this case, it was the air traffic controllers – not the pilots – who were charged with negligently endangering lives in the Icelandic airspace under their control by allowing the two planes to come within 50 feet of each other. Both pilots had filed an official airmiss report. Although the court rather oddly suspended judgment for five years, the controllers were ordered to pay the costs of a six-month trial.

The International Federation of Air Traffic Controllers, naturally alarmed by the use of criminal charges against the men, saw this as a dangerous precedent and a threat to safety. It revealed that a number of similar actions were pending in Europe. As a British spokesman pointed out, human nature being what it is, people will be less likely to report something if they believe that a prosecution may follow or if they are unable to remain anonymous. Clearly it would encourage

44

the suppression of the vital knowledge needed to prevent recurrences. Beyond its silencing effect on the mandatory reporting of airmisses, they feared that the CHIRP confidential reporting system, now financed by the Civil Aviation Authority, may be brought to an end because the anonymity of its contributors could be broken and their identity revealed by subpoena from the criminal courts. This would indeed be a tragedy for the future of air safety.

Even with the benefit of CHIRP and its US equivalent, it would be absurd to claim that all is known about every flight. There are other sources. The 'black box' accident recorder (in fact painted day-glow orange) is fed from two systems. One is the eavesdropping cockpit voice recorder (CVR) that is automatically wiped clean every 30 minutes and, routinely, after every flight unless there are specific reasons for keeping the record. The occasions on which it must not be wiped are few and are rightly jealously guarded by aircrew and their pilot organizations. The content may only be used for safety and not disciplinary purpose. The other source is the flight data recorder – usually stowed in a safe place in the tail-end – which traces the main operating events of the flight, such as altitude, speed, attitude of plane and so on. The main ingredients of this data allow the three-dimensional flight path to be reconstructed in time and space.

These and other data recorded on the ground by air traffic and radar controllers reveal a wealth of operating events from each flight to provide a valuable material for statistical, engineering and indeed safety purposes. But the amount of human detail that is allowed to drain through the official filters and restraints is limited to very specific purposes, unless of course an actual accident requires them to be fully released. There is a reservoir of private knowledge about day-to-day events – and particularly those most valuable cases of near-accidents caused by human error – that, apart from CHIRP, seldom finds its way into official archives. These, of course, are mainly restricted to statistics for analysis, accident reports and the product of the monitoring duties by the authorities. One of these sources is the system of Mandatory Occurrence Reports (MORs) – a specific list of serious events which must be reported by airlines if and when they occur.

In theory, MORs should overlap the functions of CHIRP to some extent, but the system has fallen under growing scepticism because it is widely thought that only the minimum of reports are sent in – that is, when it becomes unavoidable or the events are otherwise discoverable – and the number of MOR reports have dropped markedly since the rule was introduced. Although the content of MORs is said to be confidential, compulsory systems do not induce people to bare their souls however high their motives may be, and MORs are never likely to plumb the depths of private knowledge from the flight deck that are so vital to future safety. They may reflect flaws in the hardwear, but where there is an element of self-blame involved, silence is a human enough response. This was enough to cause the FAA in the US, which initiated a similar compulsory incident-reporting rule, to switch to a voluntary one. In Britain, the CAA has yet to respond to this cue.

This mass of recorded history from written sources or data is supplemented, of course, by a wealth of anecdote and personal experience that is less easy to evaluate. Apart from the menace of bar-propping raconteurs and their kind, no one who wanders in the purlieus of aviation can fail to pick up a deal of insight into the real human condition on the flight deck – whether it is gained at first hand or on the ground. The sum of such experiences, which cannot be rated as more than a personal hunch, is that far more flights encounter the edges of danger than are ever dreamt of in the philosophies of the travelling public: and more than many desk-bound bureaucrats and the air travel industry itself are willing to concede.

In his inaugural speech in Washington in July 1987, the FAA's new boss, T. Allan McArtor, told the airlines that passengers were clearly 'asking you to do better'. This came on top of a recent clamour of rising dissatisfaction with delays and discomforts, backed by alarm over an astonishing series of near-accidents to US airliners. 'There is,' he declared, 'a glaring need for the public to be better informed about the actions being taken to improve the system.' He promised that measures would be taken to demonstrate immediate progress 'to a doubting public'.

That in itself seems to be good enough authority for giving the passengers in the cabin a wider understanding of the problems up front on the flight deck that govern their daily destinies. McArtor's tough talk is clearly a response to a well informed opinion in the USA. Some might say too well informed, judged by the media's nationwide shrieks of alarm over the recent series of mishaps to one major and highly regarded airline ('HIGH ANXIETY AND RAGE' shrills a *Time* headline).

Delta Air Lines had won official acclaim in the US as the carrier with the best record for safety and reliability, before it suddenly fell from grace after a string of six black marks within two weeks. The story is too well known to detail at length, and if it proves anything it may be no more than that multiple coincidences are not as remote as the theory of probability might suggest (it seems that lightning can strike the ego of the victim not only twice, but six times in the same painful spot). One Delta crew shut off both engines by mistake in flight, another just landed at the wrong airport. Then followed a near-miss over the Atlantic, followed by navigational wandering on the same flight; another plane landed on the wrong runway at Boston and finally at Nashville, Tennessee, a Delta crew acted on an air traffic instruction addressed to another plane. The media added gravy to the stories with allegations of conspiracies and cover-ups – after the near-miss on the Atlantic flight a crew member reportedly whispered to passengers to keep what they had seen a secret.

The FAA announced an immediate investigation of the airline's pilot training methods, and there is no doubt that each of the incidents was serious. Time will tell whether they were causally linked or whether the concatenation of mishaps can be put down to sheer chance. But the row demonstrates not only a fully alert and informative press, but a much tougher response by the US authorities than is ever shown by their counterparts in Britain or its European neighbours. McArtor's style is uniquely American.

His public address put every carrier on notice – 'if you do not comply with your obligations to maintain your fleets and fulfill the obligations of your operating certificates, you will not operate in national airspace' – and he added a similar warning to pilots who showed themselves to be incompetent. The press reported a shady tail-piece to the story of the Delta airmiss over the Atlantic.

Over mid-Atlantic on 8 July 1987, Delta's Lockheed 1011 flying from London Gatwick to Cincinnati strayed 60 miles off its course and passed less than 100 feet below a Continental Boeing 747 en route from Gatwick to Newark. Immediately after the near-collision between the two jumbos – so close that the Delta plane was bumped by the turbulence from the other – the two shaken pilots spoke by radio and conspired to keep quiet about it. This was overheard and discussed by more than twenty other aircrews flying in the vicinity, none of whom obeyed the rules by reporting the airmiss to air traffic control. (Airliners carry short-range radios (VHF) and long-range sets (HF), so that they could talk amongst themselves on VHF out of earshot of the controllers who use HF channels to guide Atlantic traffic.)

Some of these other pilots encouraged the Delta crew's plan not to report the event but, possibly because passengers had witnessed the airmiss through the cabin windows, the Continental crew later decided to report it to their management. Meanwhile the Delta plane continued to stray until it was 80 miles off course, before the crew silently regained its assigned flight path.

The inquiry by the Canadian Aviation Safety Board, in whose area the airmiss occurred, reported two years later that the Delta crew had failed to follow established cross-checking procedures that would have revealed its navigational blunders: a disparity of 16 minutes between the planned flight-log times between waypoints and the times shown by the flight management computer in the cockput went unnoticed.

But the Safety Board was more alarmed at the astonishing cover-up by so many crews. Besides the rule that airmisses must be reported, a straying aircraft remains a danger to all until air traffic is warned – the more so when traffic is beyond the range of radar, as it was in mid-Atlantic. The full story came to light because one of the crews involved had recorded all the interchanges between the planes.

The breaking of Delta's secret lends substance to the claim that pilots can and do agree to keep silent about airmisses. British air traffic controllers claim that this is the reason why the official figures for airmisses appear to be decreasing and that the real total is unknown and may be rising. Whether or not an airmiss is caused by a pilot or his controller, they claim that – over the discrete radio channel between the two of them – there is a mutual temptation to get each other off the hook and forget the incident.

Narcotics and drugs as a threat to air safety

Widespread concern about the growing use of narcotics and illicit drugs in all walks of life, both in the United States and elsewhere, must pose a question for aviation as probably the most vulnerable theatre of any activity. While there is abundant medical literature that is relevant to aviation, there are surprisingly few hard facts about the proven use of drugs by pilots and air traffic controllers in the course of their duties. The following notes do no more than mark references to the problems that have cropped up in accident reports and other records. It is for those qualified in the relevant disciplines to assess the likely impact of a subject that is sensitive enough to suggest that too little has yet been willingly said about it in open court.

On the night of 19 January 1988, a Trans-Colorado Airlines 30-seater commuter plane crashed on its approach to Durango's airport in Colorado, USA, killing 9 of the 17 people on board. Both its pilots died. Accident investigators found that its captain, Stephen Silver, was medically unqualified to serve as a crew member on the flight due to his use of cocaine before the accident. While his co-pilot made a dangerously fast and erratic instrument approach, Silver was unable to monitor his progress or prevent him from flying into the ground five miles short of Durango's runway.

The background to the crash reveals the difficulties in detecting and proving the use of drugs by aircrew, and of even identifying addicts. In this rare case, the National Transportation Safety Board succeeded in doing so and this gives it a special importance. How this came about, and the Board's comments on the symptoms and effects of drugs on pilots, are told in an illuminating report published in February 1989.

It began with a curious story told by one of Silver's fellow-pilots who contacted the Board after the crash. Some months before the crash, he said, he had met a lady who had travelled with Silver as his wife on a passenger flight: in fact, she was not his wife but had been living with Silver. After the accident in which Silver died, the informant had contacted the lady again. She told him, 'I'm sure glad that we were able to bury him right after the accident, because the night before we had done a bag of cocaine. . . .' She was worried that an autopsy would reveal traces of the drug, and admitted that she and Silver had used cocaine periodically.

The Safety Board tried to contact the lady but her attorney refused to permit this and denied that she had been with Silver the night before the accident. The Board then contacted another of Silver's girlfriends, who said that she 'knew right off there was some kind of drug problem' and she confirmed that Silver and his later partner were taking cocaine together. An autopsy was then performed on Silver, revealing that the amount of cocaine he had taken the night before his death was 'likely to have degraded his piloting skills' and that this had contributed to the accident.* In its report, the Board stresses that the use of cocaine is increasing in the United States and, as this accident demonstrated, 'its use by pilots poses a threat to the safety of the flying public'.

The difficulty of identifying drug-users presents a daunting prospect. Even Silver's medical examiner had found no indications of it at his last mandatory health check, and the Board only established the truth after ferreting deeply into his private life. Even those who rub shoulders daily with cocaine addicts may not notice any symptoms: the outward effects can be subtle as they can depend on a number of factors such as methods of ingestion, tolerance and the interaction of individual factors that may mask physiological responses.

The investigators also found that the complexity of the effects of cocaine makes them hard to distinguish from impairment caused by a host of both licit and illicit drugs. Finally, the Safety Board emphasizes the difficulty faced by the aviation community in tracking down and controlling the use of cocaine in the cockpit. It urges the Federal Aviation Authority to undertake immediate research towards a better understanding of all the drugs that impact upon pilot performance, and that the findings should be 'actively disseminated'.

In Britain, the Civil Aviation Authority reports that as far as it is aware, there have been no problems with flight crew incapacitation due to drugs and none have been identified through regular CAA medical examinations, which include blood tests.

In random drug testing undertaken in 1989 by the US Department of Transportation, 57 air traffic control staff employed by the Federal Aviation Administration were identified as users. Out of a total of 50,000 government workers tested, 0.5 per cent were positive (mostly for marijuana which is easily detected in urine for several

_____
* Blood sample: 22 nanograms/millilitre of benzoylecgonine. Urine sample: 22 ng/ml of cocaine and 1,800 ng/ml of benzoylecgonine.

weeks after use). Two of the controllers were fired after a second test, 21 were rehabilitated and returned to work, while the remainder are undergoing rehabilitation. Any government employee whose test is positive can be fired.

3 | The men who fail the pilots

Nobody will be surprised that pilots and their crews account for most of the accidents in the air. Day after day flight crews are called upon to maintain a uniquely high level of professional skill for many hours at a stretch. A single lapse can end up as another accident statistic. Every action and word is recorded and can be played back after each flight whether or not there is an accident – the crash-proof 'black box' flight recorder will preserve the evidence. It adds up to a high degree of stress that is unmatched even by other professionals – such as hospital surgeons or road and rail drivers – who require the same kind of attentive skills and for whom a momentary lapse can also be fatal, but who do not have to contend with the force of gravity into the bargain.

Every flier is backed by a legion of non-airborne specialists spread through the airline, airport staff, air traffic controllers and a hierarchy of chairborne national and international regulators, all sharing the primary aim of safety in the air. None of these battalions of people who serve the airline industry from the ground is subject to such constant scrutiny.

International regulation

Bureaucratic control is topped by the International Civil Aviation Organization (ICAO), based in Geneva and Canada. Like all world councils it faces the diplomatic constraints imposed by the need for unanimity among all its member states, and this tends to slow the pace to that of the least co-operative member. But it has laid down a set of rules that by and large have won acceptance and shape the internal laws of each state.

At this national level America and Britain share the same approach of two-tier control. The law-making and policing body in the US, the Federal Aviation Administration (FAA), is entirely separated from the National Transportation Safety Board (NTSB), whose role is to investigate accidents and extract the lessons to be learnt from them. These recommendations are submitted to the FAA for executive action. This set-up succeeds in maintaining a useful amount of creative tension between the fault-finders and those responsible for preventing them. The Safety Board scrutinizes the whole context of each accident and its final conclusions come without fear or favour. As often as not the Board can be seen to be barking at the heels of the FAA itself.

The Civil Aviation Authority is the FAA's counterpart in Britain, balanced by the Department of Trade's Air Accident Investigation Branch at Farnborough which mirrors the roll of the NTSB in the USA. There are differences of emphasis between the two national systems that need to be considered, but they share a principle that clearly stands as a model for many other countries to emulate. No human system can be perfect or without kinks and shortcomings, and these reveal themselves clearly enough in the annals of air safety.

The airlines themselves are the last line of defence which must operate the safety regulations handed down to them from above as a condition of their licence to fly passengers. Their training, flying and maintenance standards are of course closely controlled, as well as the integrity of the aircraft they operate. Many airlines go well beyond the safety limits set by law, but we shall see that there are a few who do not even reach them.

For reasons of staff economies, both the FAA and the CAA use the system of appointing their local representatives from the ranks of airlines' staff. This two-hatted role is a practical expedient, but it hardly diminishes the obvious risk of a too cosy relationship between the airline and an employee with a government badge on his lapel. It has led to some of the sharpest criticisms from accident investigators. And there is a good deal of concern that recent cost-cutting by governments has reduced full-time policing staff, so increasing the need to farm out more responsibility to airline staff.

Paris DC-10 crash: human error beyond the flight deck

The need for vigilance about such a potential weakness got a mighty boost in the aftermath of the Paris DC-10 crash in March 1974, which killed 345 people and rated as the world's worst air disaster at the time. A faulty cargo door had caused sudden cabin decompression and ruptured the aircraft's flying controls. That catastrophe has become such a legend and a benchmark in the pursuit of air safety that it must inevitably be drawn into any purposeful discussion on the subject.

The DC-10's Turkish pilots were blameless. One reason for the faulty door was the still unexplained action of a works inspector – an employee holding an FAA local inspector's licence – who certified that a safety fix for the door had been made to the plane, when it had not. But the inquiry showed that this was but one fault among many. The full history of laxity, duplicity and negligence that led to the dismemberment of the jumbo plane and its human cargo in the charred forest at Ermenonville, in the outskirts of Paris, only emerged in the court rooms years after the event. A 'carbon-copy' decompression caused by the same cargo door on another DC-10 had happened two years before over Windsor in Ontario, Canada. The lessons were learnt but pushed aside.

Although that saga has since been copiously documented,* sadly its lessons have not yet been well enough learnt as a parable of lasting importance for the aviation industry. The ten years and more since that story became known have dimmed a message that deserves to be recalled.

Fifteen days after that first scare over Canada, Dan Applegate, the director of product engineering with Convair, which was the leading sub-contractor for the manufacturers of the DC-10, had expressed his alarm. In a now celebrated memorandum to Convair's vice-president dealing with the DC-10 contract, he wrote that unless immediate steps were taken to cure 'a fundamental failure mode', he believed that in the years ahead it was inevitable that 'the DC-10 cargo doors will come open again and I would expect this usually to result in the loss of the airplane'. Those steps were not taken, and what is worse, his warnings never reached the FAA, which had certified the plane as fit to fly months before the Canadian incident.

There was more. A year before Applegate's warning, factory tests had shown that the explosive decompression caused by the loss of the door in flight would lead to catastrophic structural failures – the

---

Destination Disaster, *Sunday Times* Insight Team (Granada/Hart-Davis, MacGibbon, London; Quadrangle/NYT New York, 1976).

54

collapse of the cabin floor and the severing of the vital flight control lines routed underneath it. Despite this, the response of McDonnell Douglas to the Paris crash was to blame it on a baggage-handler at Orly airport who, it claimed, had failed to close the cargo door properly. Yet the gimcrack design of the door-locking mechanism should have been known since before the Canadian failure and seen to be flimsy and lethal. This included the cockpit warning light that should have warned the crew that the door was not secure.

In the aftermath of the Canadian flight, its American Airlines captain, Bryce McCormick, tersely advised McDonnell Douglas to 'get that door fixed' and got the response: 'Mac, that's a promise.' Ten months later the door blew off again and killed all 345 on board the second DC-10.

In Britain, the news caused one airline to respond in a more practical way. Freddie Laker, of Laker Airways and Skytrain fame, sent his men down to the hangars to fit stout bolts to secure the rear cargo doors of his fleet of DC-10s without waiting for formal orders which were to come over the wires later. In the USA, McDD later admitted that there had been a few minor problems with the doors in the past. The investigations that followed, however, exposed records of approximately 100 instances of doors failing to close properly.

Neither did the FAA come out of the Paris inquiry untarnished. In its role of scrutineer it had preferred a cosy gentlemen's agreement in handling the company's shortcomings, rather than using its legal powers to check them. After the Windsor episode, FAA officials allowed McDD to deal with the problem through its own channels and the company issued three advisory service bulletins to its DC-10 operators. These specified modifications to the design of the door-locking mechanism, one of these being the fitting of a metal support-plate behind the handle to prevent the complex interior linkage from bending. There is bound to be some time-lag before recipient airlines can deal with the routine flow of service bulletins, and neither Laker nor Turkish Airlines – nor an unknown number of others – had fitted the support-plate to their doors when Paris happened. Had the FAA raised the status of the modifications to that of a mandatory Airworthiness Directive or AD, the planes could have been grounded until the safety changes had been made.

It was this softly-softly handling of McDD that led to what must be one of the harshest verdicts on the conduct of the FAA. A report by the House of Representatives in 1975 concluded that 'through regulatory non-feasance, thousands of lives were unjustifiably put at risk'. How, one wonders, would the statisticians now express the risk of flying on all those DC-10s before the day when Freddie Laker bolted up his baggage doors?

Those risks – and all those embodied in the DC-10 – were known and present before the London-bound Turkish airliner lifted off from Orly airport at 12.30 p.m. on that fatal Sunday. But they were not known to the 334 doomed passengers who had climbed aboard, many of whom had been switched from a British Airways flight that had been cancelled after delays caused by an air traffic controllers' strike in London. Nor did its aircrew have an inkling of the latent dangers that were about to burst upon them a few minutes later, although it had all happened before. Their last few words replayed from the cockpit voice recorder, later discovered from the debris of the crash, told how they had fought vainly to trace the cause of the unresponding flight controls of their paralysed plane in the last seconds before it hit the ground at 497 miles an hour.

The French authorities were still mystified when, some hours later, I stood amongst the ghastly carnage at the site of the crash. Some said there had been a bomb on board, others talked of a mid-air collision. None of them then knew that the secret lay buried in the records of Bryce McCormick's flight over Canada.

Young trees are now pushing up in a strangely open swathe cut through the centre of the dense forest of Ermenonville. The saplings in Dammartin Grove, as it is known, are beginning to obscure a memorial erected there to the hundreds of lives lost in that accident. With hindsight, however, to describe it as an accident looks like less than the whole truth. A more apt epitaph might be that they were killed by a tragic chain of human failures that had been hidden from sight too well and for too long.

Nobody can say how many accidents have been averted by conscientious and foresighted people in the industry, but when a mishap beyond the flight deck does occur the record shows only too often that safety regulations have been breached or flouted. It is hard to resist the conclusion that accident reports are revealing only the tip of an iceberg.

Flight 901

Let us pick another case at random from the list of recent fatal crashes, a case that did not even rate a mention in one leading US daily – the *Washington Post* – and seems to have gone largely unnoticed elsewhere, in contrast to the vast publicity lavished on the notorious Paris DC-10 affair: yet the similarities are unsettling.

Tragedy struck among the Caribbean islands early on a perfect summer's day when a small commuter plane, bound for the nearby Virgin Islands, crashed into the sea off Puerto Rico in August 1984. Any one of us could have found ourselves aboard such a plane holidaying around the surrounding islands: you can't visit them easily any other way. Its pilot and all his eight passengers were killed.

The airline, Vieques Air Link (VAL), operated a busy inter-island service with its nine light planes flying some 40 scheduled services a day. Flight 901 was listed to leave Vieques airport at 8 a.m. for the 30-minute trip to St Croix in the US Virgin Islands. There was a swift turn-around of light planes dropping in and out to and from other island airstrips and a Britten-Norman Islander, a rugged British twin-engined 12-seater, was detailed for the flight. In the good weather prevailing, only one pilot was needed. He was 21-year-old Captain Miguel Garcia.

Earlier that morning after flying a short return trip, he had taxied his plane to the pumps for refuelling before his eight passengers climbed aboard with their luggage. Shortly before 8 a.m. his Islander buzzed down the runway and lifted off over the shoreline. As it reached 200 feet above the sea his left-hand engine choked and stopped. People on the airport saw the plane's nose go up, the plane yawed sideways until it started to lose height. Then it banked abruptly to its left and the nose dropped into an uncontrolled plunge into the sea, half a mile from the end of the runway.

A US navy helicopter, with swimmers and divers aboard, was the first to reach the scene 37 minutes later but they found no survivors. All had died strapped into the seats of the submerged plane. Four, including Garcia, had died from impact, five were drowned.

The fatal flight lasted no longer than it takes to read these bare facts. But the lengthy post-mortem into its cause revealed a bundle of contributory factors. It took the top off a grisly story of incompetence, negligence and worse besides, that had combined to cause these needless deaths. As nearly always, the accident had multiple causes, and to grasp the essential warnings from such a crash requires patient attention to the details.

The airline, its planes and pilots were all properly accredited by the FAA for scheduled passenger operations. If it was not a sheer accident, what therefore could have gone wrong?

In a properly ordered flight, a pilot should be confident and able to cope with the loss of an engine, even at the most critical time just after take-off. Specific drills are laid down headed 'Failure of One Engine After Take-Off – Immediate Action' and all properly trained pilots are so practised in them that they become second nature. Periodic flight checks ensure that they keep them smart. As the Electra mishap at Reno proved, the seconds after take-off are the most fraught, but if the right actions are done rapidly and precisely, all multi-engine airplanes must be capable of continued safe flight.

The Islander is no exception, although piston-engined planes can have smaller margins of recovery power than jets. The pilot's overriding need is to gain a critical airspeed at which the plane is certified able to continue a gradual climb on the remaining engine. Light planes like the Islander will not do this unless they are 'cleaned up' by reducing drag from the landing flaps and by feathering the blades of the useless propeller.

The moment an engine fails, the sequence goes something like this on a Islander. Full power applied to both engines – instantly. Identify failed engine (not so easy under stress – many have shut down their remaining good engine from panic). Start the close-down of the culprit engine and feather its propeller. Check power from remaining generator on the good engine. Allow the airspeed to build up to 75 m.p.h. (the optimum climb-speed for single-engine flight: this may need a gentle controlled descent even at such a low height). Retract landing flaps. Cut fuel to dead engine. Once assured of the safety speed, there follows a string of tidy up actions. Novices are told, proverbially, to jump to it like a one-armed paper-hanger while the other hand – and remaining part of the brain – flies the plane, in other words, keeps it from going out of control.

In the climb-out from take-off the sudden loss of half the available power will quickly bring the plane teetering near the point of stalling speed – the moment at which it goes out of control and drops out of

58

the sky – unless properly handled (see pp. 27–9). Once its safe single-engine flying speed is attained, only then comes the time for the pilot to think about a careful turnabout back to the airport. The Islander can safely accomplish all this, and no pilot who has not demonstrated that he can safely coax her home should ever be allowed at the controls.

Young Miguel Garcia claimed 400 hours of twin-engine experience, but in fact he had only 145. Nor did his total flying time exceed 1,000 hours as he alleged. With only a commercial and not a full airline pilot licence, he was unqualified to fly a scheduled passenger flight – and for reasons which will become only too apparent. When the dreaded moment struck, he fumbled it badly. Flaps were not retracted, the dead propeller remained unfeathered, and he allowed the speed to decay even below the remarkably slow flying limit – 45 m.p.h. – of the safe old Islander. Predictably, she nosed down and over into an uncontrolled stall.

In fact it seems that Garcia failed to initiate any of the emergency procedures. But there were mitigating circumstances. It was discovered later that his plane was more than 600 lbs overweight and his load was badly balanced – both of which would mean that he had an even shorter response time available to complete the emergency actions than he should have had. These factors compounded his difficulties, and incorrect loading was found to have contributed to the crash that followed.

The engine had failed because there was water in the fuel tank from which it was fed. After refuelling on the ground, Garcia should have obeyed the rules by waiting for any water to settle at the bottom of the tanks, before routinely draining off a sample and checking it. No one saw him doing this chore and even if he had, there was not enough time for any water to settle between fuelling and take-off.

The airport fuel tanks also contained water well above the accepted safety limit. It had rained heavily during the night, and that morning's regular sampling check had shown the tanks to be dangerously contaminated. They should have been pumped clear before use, but the warning was ignored and nothing was done. And some very odd things were done to the tank *after* the accident.

A strange drama was played out around the Vieques airport buildings that sunny August morning. At 08.05 the duty man in charge saw the plane crash and dropped his mop to run towards the beach. When he was halfway down the runway, he saw another plane had landed and would need to be refuelled. He stopped his dash towards the stricken plane and returned to the pumps, where he refuelled the arrival as though nothing had happened.

Whatever allowances are made for the fact that priorities can get mixed up in a panic, the later history of the crash suggests something else. Might it have been that management imperatives for a quick turn-around between flights had been so effectively ingrained in the work staff that it had become an instinctive priority even above a crashed airplane? Those who know the pressures on a small commercial airline will remember the haunting cry to 'keep those planes in the air', or else.

By 09.05 police and fire chiefs arrived on the scene. They walked from the main hangar at 09.30 and observed pumping activity at the fuel storage tanks. Two airport men were pumping out liquid and dumping it over the concrete ramp. The airline president, who had joined them, cupped his hands into the liquid and said: 'Hey, this is water and mud!' Safety Board investigators later commented drily that no explanation was offered as to why the tank was pumped out about an hour after the accident, beyond preventing more suspect fuel being supplied. Clearly all operations should have been halted after the crash. Tampering with the storage tanks before the officials arrived raises obvious implications.

By 10 a.m. the airline's insurance broker arrived and, with the president, drained fuel from the tanks of another plane standing on the ramp. Each tank is fitted for this purpose with a drain-tap placed at its lowest point under the wing. The liquid that came out was water, although the pilot of the plane said that he had drained his tanks after refuelling that morning. Some days later when the investigators checked the filter element and fuel bowl fitted to the pumps, it was found to be clean and new. Four days after the accident they learnt that it had been changed on the day after the accident.

There was still worse to come from the airline's paperwork. Flight 901 scheduled for 08.00 that day had been 'cancelled' and replaced by an 'on-demand' charter flight, quickly re-designated 901A. Garcia's limited qualifications allowed him and others like him to fly on-demand but not scheduled flights. The Safety Board concluded that his flight was in fact a scheduled flight and that the bogus on-demand label had been common practice in the airline. It had 'circumvented the intent of the regulations prescribing a high standard of safety for commuter airlines who provided scheduled services to the traveling public'.

A deliberate motive, the Board concluded, had been to gain economic advantage by using non-airline pilots: the use of the inexperienced Garcia had contributed directly to the cause of the accident. Its verdict was that management had shown a flagrant and reckless disregard for Federal weight and balance procedures. There had been merely 'a token effort on the part of those involved to comply with requirements on paper and not in practice'.

Nor did it stop there. The ditching crash procedures were inadequate, and five out of ten life-jackets were found leaking or faulty. Garcia's pilot training with the airline was also inadequate, showing only 53 hours' flight experience on the Islander. The airline official responsible for calculating the loading of the plane under-guessed the weights of the passengers and their baggage when they should have been properly weighed. He had entered an inexplicably low figure of 120 lbs in the flight manifest for cargo carried, and when questioned about it he said he could only remember a box of samples. Divers at the wreckage later found hundreds of mangoes on the ocean floor spilled from four large containers and close to them were six suitcases which alone weighed 300 lbs.

The Safety Board wound up by slamming the FAA's surveillance of the airline as ineffective and a contributory cause of the accident. The FAA's local inspectorate should have detected a host of violations of the regulations among other deficiencies and irregularities: among the 73 inspections it had made, it had reported only three as unsatisfactory. The Board went as far as to say that the airline's persistent use of unqualified pilots strongly suggested that the FAA had condoned it and, but for this, 'the accident might not have occurred'.

Ten months after the crash the FAA suspended the airline's operations on the grounds that the staff had knowingly prepared false manifests for the Safety Board, had violated load regulations and finally because of 'VAL's careless and reckless operational behaviour which endangered the lives and property of others'.

4 | Acts of God?

'In the unlikely event . . .'
(Start of the air hostess's safety briefing on the Jakarta
flight, 24 June 1982)

With an uncharacteristically biblical turn of phrase, insurance men like to file fluke or freak events under the heading Acts of God. The phrase is, of course, near to the hearts of the travel industry, which would have us believe that mishaps lie in the lap of the gods rather than under their own control.

Whatever an act of God may or may not be, in the realm of aviation some freak events do seem to fall under that heading – at least at first sight. Consider the fate of the captain of an Indian Airlines DC3 who was killed when a vulture burst through his cockpit windscreen. His co-pilot took control and made a safe landing. Many years later in the jet era, the captain of a British Trident was luckier when a 20-lb. vulture crashed in and landed in his lap while he was flying over Lahore, India, in 1970. The mess of blood and feathers must have been appalling, but on that occasion it was only the kamikaze fowl that died. The pilot landed without further mishap. More recently in 1982 a 4-lb. duck killed the pilot of a Learjet in the States.

Flight Lieutenant Stephen Fox was luckier by a hair's breadth when a buzzard hit his Royal Air Force Harrier during a training mission over the north German plains in July 1987. The impact at 500 m.p.h. knocked out most of his control systems and damaged the jet engines, besides covering him in blood and feathers. The bang triggered the compressed air bottle that lowers the wheels after an emergency loss of main power, and down they came, although the undercarriage is designed to be extended safely only at speeds below 290 m.p.h. This threw his plane into a terrifying and battering spasm of deceleration.

Even worse, the emergency air power will only keep the wheels locked down for five minutes. Despite his smashed instruments, he managed to pinpoint the nearest German airbase and fought the crippled Harrier down with less than a minute to spare – without landing flaps to slow him up and having to guess his speed and height. He had chosen not to eject safely by parachute and his squadron commander called his effort 'brilliant real seat-of-the-pants stuff', for which Fox was awarded the highest peacetime honour, the Air Force Cross.

More recently, in January 1988, the crew of an Ethiopian Boeing 727 smacked into an eagle during the most critical phase of the flight – the approach to land at Khartoum airport in the Sudan. The bird struck the nose radome and entered the cockpit, tearing out the co-pilot's rudder pedal and severely injuring his left leg. Debris damaged the third rear engine in the tail of the plane, but they landed safely without further injuries.

'Jakarta, Jakarta, Mayday, Mayday'

But the category of freak happenings seems to enter a shaded area with the celebrated story of the British Airways Boeing 747 that experienced a traumatic event while cruising at 37,000 feet over Jakarta in 1982. Minutes before it happened, Captain Eric Moody had left his seat in the four-engined jumbo to visit the crew toilet but, finding it occupied, he strayed into the first-class area to chat with a stewardess. The flight from Kuala Lumpur was more than half-full, with 247 passengers on board plus 91 tons of fuel for their long trip to Perth, Australia.

Suddenly Moody got an urgent call to return to the flight deck, and as he jumped up the stairs he noticed puffs of what he thought was smoke billowing out of vents at floor level that brought an acrid, electrical smell. When he rushed into the flight deck he saw the windscreens ablaze with an intense display of St Elmo's fire – in smaller quantities this is a well known and rather unnerving flickering phenomenon caused by flying through an electrically charged atmosphere. As he strapped himself into his seat, he looked at his weather radar but saw nothing significant. His co-pilot, Roger Greaves, was pointing out of the windows at the engine intakes which were glowing as if they were lit from inside and – even more mysteriously – the big front fans appeared to be turning backwards, apparently due to St Elmo's stroboscopic effect.

Now everything happened fast. Through the windscreen they saw what seemed to be a hail of tracer bullets and, more alarmingly, the smoke that had got into the air conditioning system was building up a dense smog in the cockpit. Before Moody could speak, Barry, the flight engineer called out 'Engine failure number four!' (the outer right-hand jet). Now more alert with the adrenaline flowing, the crew snapped into emergency drills as they tried to reason out the cause. Then came Barry's voice again:

'Engine failure number two . . . three's gone . . . they've ALL gone!'

Captain Moody stared at his instruments in disbelief. Four engines simply do not fail, his training told him. The dials spelt confusion because the onset of an electrical power loss froze some of them and the needles dropped off the scales on others. But the airspeed indicator still functioned and the speed was dropping fast. Moody was forced to accept the unbelievable – the jumbo had become a powerless glider. It was an unprecedented situation, but with inbred calm he told the co-pilot: 'OK Roger, put out a Mayday.'

His mate pressed the transmit button on his control stick and complied: 'Jakarta, Jakarta, Mayday, Mayday Speedbird 9. We've lost all four engines. We're leaving 370 [37,000 feet].' Meanwhile the whole crew fell to working through the emergency checklist to seek the vital clues to their predicament. As they said later, they were stricken by that familiar gut-stirring question, 'What have we cocked-up?' Such thoughts did not hinder the job of getting the engines – or at least some of them – back into life. But so far they hadn't responded.

As they descended through 26,000 feet the cabin pressurization system warned a failure. The crew reached for their oxygen masks, but Roger's fell to pieces in his hand. Moody faced the choice of a continued slow descent that would leave Roger exposed to the debilitating effects of anoxia (oxygen starvation), or losing valuable height by plunging into an emergency descent to the survivable atmosphere at 12 to 10 thousand feet below. As the jumbo turned back towards Jakarta, he ordered the maximum rate of descent. Emergency batteries lit the cockpit, but without generator power from the four dead engines their navigation systems went adrift.

They no longer knew their exact position above the 11,000 foot peaks of the mountain range below them.

As they fell through 20,000 feet Moody noticed that Roger, with the skill of desperation, had managed to assemble and fit his mask despite his emergency work-load. Airspeed indications, vital to maintaining control, now became suspect. Unsuccessful attempts to re-light the engines only resulted in igniting the fuel in the slip-stream behind them and so giving the passengers the impression of all the engines being on fire. Their oxygen masks had already deployed in the cabin, but Moody had to find time to speak to them.

'Good evening ladies and gentlemen. This is your captain speaking. We have a small problem. All four engines have stopped. We are doing our damndest to get them going again. I trust you are not in too much distress.'

He called the senior flight attendant on to the flight deck to explain the state of affairs but the steward couldn't understand the captain through his oxygen mask. In any case, a glance round the distracted crew told him that he was clearly *de trop*, so he nodded and went back to help his passengers.

Very naturally, there was fear and alarm in the passenger cabin, but it never became panic. As the jumbo's strange, silent descent continued, the extremity of the emergency became only too plain and it brought that strange stoicism that can come in the face of a common peril. The drama inside the passenger cabin of the stricken plane is graphically described at first hand by one who experienced it herself, Betty Tootell.* Her book's pages reveal one poignant sidelight that touches a central theme of ours here with a gleam of irony. She tells of an exchange between one lady who tried to comfort her distressed companion by reminding her that the statistics about air travel showed it to be much safer than travelling by road. 'But,' the author reports, 'the passenger remained unconvinced.'

Moody now faced the novel prospect of ditching a powerless jumbo onto the sea in the darkness. He aimed to track back over the ocean and away from the mountains. By a stroke of fortune, he had practised four-engine failure drills on a flight simulator only months before. Real 'dead-stick' landings – with no engines – are things that trainee pilots may practise in light planes, but over land and by day.

* *All Four Engines Have Failed*, Pan Books, 1986.

With the rapid imagery that comes under stress, a childhood memory came to him of a trip with his father to watch flying boats landing on the sea. He knew that they couldn't do it at night because of the difficulty of judging height above water in the dark. . . .

By now the plane had been gliding helplessly for thirteen minutes. His thoughts about the nightmarish prospect that seemed about to become a reality within the next six or seven minutes were cut short by shouts of glee from the crew as number four engine started up. Power surged back into the plane's systems and in the next 90 seconds the other three engines came alive in time to check their descent at 12,000 feet. Nursing the jets carefully, they climbed to a safe 15,000 feet above the high ground. Number two engine then misbehaved and had to be shut down; oddly enough, they still suspected that some mistake or oversight had caused the engine failures.

No wonder they were mystified. The events of the last few minutes, unique in aviation history, provided the fewest clues. The only difference between the view from the cockpit before and after it all happened was that the stars could no longer be seen – the normal result of flying into cloud. The real cause was to reveal itself half an hour later as they made an emergency approach to Jakarta airport.

They were surprised when they couldn't see the ground or the runway lights clearly. Then they spotted lights through the co-pilot's side-window, but when Moody lined up with the runway the lights disappeared again. Soon they realized that the front windscreen was almost opaque. Moody continued by glimpses through his side-window as the crew monitored height and distance from the instruments. When they came near to the runway, a diffused glare floodlit the cockpit. This was comforting because it meant that they were 'in the general proximity of the runway'. But the seconds waiting for the wheels to touch the ground seemed like an age. As they smoothly kissed the ground, the passengers in the cabin broke into a thunderous chorus of cheers and clapping.

Moody braked to a stop at the end of the landing run and radioed airport control: 'Speedbird 9, I can't see with the light in my eyes. I'll hold it here.' Perhaps not a bad place to be, considering.

With the engines shut down, Moody left the captain's seat and slipped down the stairs to the lower deck to speak to the occupants of the cabin. It was then that he noticed a layer of black dust covering everything, some of its sharp and gritty. Puzzled, he climbed back to the flight deck and mentioned it to the crew. His flight engineer, Barry Townley-Freeman, had been in Jakarta a month before and something that he had read in the local press tickled his memory.

'It's volcanic ash!' he cried. 'That's what we've been flying through!'

Barry's inspiration explained everything. They had flown through a vast column of volcanic ash erupting from Mount Galunggag, a volcano on the west of the island of Java and about 100 miles to the south-east of Jakarta. Its last serious eruption had happened in 1918. But in the two months preceding the flight, there had been ample warnings of its coming crescendo on the night of 24 June that had choked the jumbo's four engines and sand-blasted the surface of its windscreen so that the crew could not see ahead. The impact with electrically-charged particles of dust also explained the lively display of St Elmo's fire.

But any post-mortems were to come later. Meantime, at the airport, it was time to celebrate a remarkable escape from the elements.

The crew, feted by the world press, won honours for their professionalism in their handling of the unknown. The passengers were indeed fortunate in their crew. Quick and cool thinking had extracted them from an ordeal that might well have led to disaster in other hands. Reflecting on the perils of the volcanic dust-cloud, Captain Moody later felt that if it had stretched down to the earth, 'I think we would have gone to ground.'

A native of Hampshire, Eric Moody seems well versed in the rustic expressions of his countrymen. When pressed by importunate reporters for a quote about how it felt to land a jumbo at night with no forward vision, he replied in exasperation: 'It's a bit like negotiating one's way up a badger's arse.' It was one description of the event that was not reported in the press.

When reports of the flight over a known volcanic region reached the press, it was natural that the world wanted to know why the pilots were not warned about a massive eruption that had been known to be in progress for at least four hours before they encountered the rising dust-cloud. Even if it had not been effectively predicted and included in the pilots' weather briefing before they took off from Kuala Lumpur, air traffic controllers have a duty to radio any significant or hazardous changes in the weather (SIGMETS) to pilots while they are en route. Captain Moody was given no such warning as he headed for the dense cloud ahead, although he was in routine radio contact with air traffic control throughout the hour-and-a-half flight from Kuala Lumpur. His crew gave normal position reports and flight details at each appointed reporting point and these were acknowledged and confirmed by the radar controllers.

The cosmic events of the Jakarta flights are worrying enough in themselves, but they reveal a glimpse of a whole chapter of dangers faced by aircrew on international routes that are seldom allowed to reach the travelling public. They might ask, how can it be that in the face of all the claims about the safety of air travel, a British Airways jumbo is allowed to flounder into a volcanic eruption with near-fatal results, and when the massive clouds it produced were clearly visible from satellite weather pictures hours before it happened? A closer look at Captain Moody's problems helps in the search for answers.

Weather data

Pilots are, of course, meticulously briefed before each flight, and an important part of this is weather information – charts of winds and temperatures at different flight levels, warnings of storms, turbulence and forecasts of conditions at destination airports. This 'Met' information now has the added accuracy and input from weather satellites, allowing instant updates by radio during flight.

Added to this, a routine part of the pre-flight briefing is a search through a list of current warnings. The most urgent of these are circulated by teletex on an international network and all are known the world over as Notices to Airmen or NOTAMS, for short. A known volcanic eruption near any route would obviously call for top-priority notice. More sudden hazards must be radioed to the pilots by SIGMET, as already noted. These are international requirements as agreed and laid down by ICAO.

In spite of all this armoury against the unknown, the systems crashed. Those responsible in the area not only failed to feed the known hazard into the NOTAM network before the flight but, incredibly, did not do so until the following afternoon and then by means of undated telex (weather data is useless and even dangerous without time and date). And this was long after the sensational landing of the damaged plane at Jakarta – the control centre for the region. Little wonder that another flight, a Singapore Airlines Boeing 747, was allowed to fly into the same area nearly three weeks later: three engines ingested ash and had to be shut down, but the pilot, now alerted by British Airways experience, took avoiding action and returned to Jakarta safely before worse befell.

Subsequent reports from a reliable source claim that the local Met men did not know what a volcanic cloud looked like on a satellite picture until they first saw it in a photograph published years later (it was taken by the US NOAA-7 satellite on the night of the Jakarta epic). One ICAO official admitted after the event that 'we just hadn't thought about it' – and in mitigation it is true that the 300-odd active volcanoes around the world, one third of them lying in Indonesia and the Philippines, happen to have been strangely quiet since the advent of wide-body jets in the late 1960s. Nonetheless, after their miraculous escape, Captain Moody and his crew learnt that the Mount Galunggag volcano had been exploding fitfully for some weeks before it broke into its first major eruption for many decades on the day of the flight. Earlier that month the Indonesian authorities had evacuated thousands of inhabitants from an area already devastated by intensified eruptions. Satellite weather pictures had shown the massive clouds from the renewed activity on 24 June before the flight had left Kuala Lumpur.

Steps have since been taken to maintain a watch on volcanic activity so that in future pilots should be warned about the probabilities of such high dust-clouds developing, especially at night. It can only be hoped that a lesson has been learnt.

There is a special need for this. Unlike storms and the dreaded cumulo-nimbus thunder clouds – which remain a threat even to high-flying jets – volcanic clouds do not snow up on a plane's weather radar screen. This will register echoes from water droplets – rain, show, hail and dense cloud – but not from dust particles. The result was that the British crew were totally unaware of reasons why the engines had failed until after the flight. In more senses than one, they were flying blind.

The ordeal of Captain Moody and his passengers lifts a veil on only one of the inadequacies that threaten the safety of international flights. There are others that seldom get a mention in the cause of maintaining good international relations between countries. Propriety and protocol demand that lapses in safety standards are not discussed in public. Nor would aircrews want to rock the boat and so risk the co-operation of the foreign air traffic and airport staffs on whom their lives so often depend.

But there must come a point at which the protection of life and limb takes precedence over such niceties, although it clearly remains prudent not to be too specific. There would be little point in maintaining the vast apparatus of safety requirements and standards as seen in the western world, if international routes from those very countries are then subjected to indifferent or dangerous handling of the same flights as they pass through foreign airspace. It is the function of ICAO to ensure uniform standards worldwide, but like all such aspirations, diplomacy gets in the way: that the speed of the convoy is the speed of the slowest ship is an analogy that holds good in most international assemblies, no less ICAO.

Due to these inhibitions, it is inevitable that much of the knowledge about foreign shortcomings is anecdotal. For instance, air traffic control in parts of Eastern Europe may have improved since it caused routine complaints. At one major tourist airport, it was reported that the controllers simply clammed up when emergencies developed and when their help was most needed. It may have been due to language difficulties, or done to avoid penalties for any mistakes recorded on air traffic tapes. In more than one case, a British pilot was left without any information about possibly conflicting traffic as he made a night approach to the airport – and without any confirmation of the vital data needed for an instrument approach. There are many other dark regions where pilots fly with not much more than a hope and a prayer. They tend to describe these poorly regulated areas as 'Indian country' and some are not far distant. To state it conversely, the countries whose control of airspace is generally held to be beyond reproach include the USA, West Germany, Holland, Australia, New Zealand and the UK.

To return again to Captain Moody's flight over Jakarta, the saga is relevant to the argument about freak and not-so-freak happenings in the air. Although it had not been foretold, the potential hazard was

70

humanly predictable. Whether or not the insurers of the damage done to the plane filed it as an act of God is not known, but if they had done so they would have a poor case in both logic and law.

Collisions with vultures and ducks add to a long chapter of dangerous bird-strikes that has led to the toughening of windscreens so that they can now withstand at least a 4-lb. bird-hit at 470 m.p.h. Theoretically, a computer analysis of the comparative densities of airspace infestation by birds and planes might calculate the probabilities of collisions between them, at a given time and place. However useless the answer might be, the possibility again reveals at least some degree of predictability. It could be argued, then, in purely absolute terms that the purely freak accident remains elusive, if not illusory.

The idea of the remote, million-to-one cause dies hard. Yet, as will be seen, accidents that are written off as freak occurrences have a nasty habit of recurring before the ink has dried on that first and faulty verdict.

Part 2
Secrecy

5 | Corporate liability

In Part 1, we have seen all the dangers of flying. In Part 2, we will discover why the public does not know more about these dangers.

In the jargon of the times, customers seeking value for money – or indeed safer products – are exhorted to shop around. In a street market this would be sound advice, and in other free-market sectors it still, of course, holds good. But a meaningful choice assumes access to facts and to the free and truthful description of products, or at least a chance to handle the goods or check out a service. Nobody needs to be told today that most of the 'facts' come from the partisan advertising and public relations fraternity whose first rubric is to expose the selling-points and conceal the flaws, so far as their skills allow them to sail up to the edge of the laws against misrepresentations and deceit. The need for truthful descriptions is as old as the English hall-marking laws that have guaranteed the quality of gold and silver goods since AD 1300.

Holiday brochures apart, few commodities come under such tight wraps as an airline ticket. As it has been said they are the least informative of documents and do not answer any of the questions you might have in mind as you shop around for the safest, cheapest, most convenient and most comfortable flight. To this extent, our ideal customer standing in the queue to buy an airline ticket is disarmed by ignorance and becomes largely a hostage to fortune (hoping, no doubt, not to be a hostage to anything worse). He has a higher claim to know more about what he is buying than when he pays for a suit of clothes or a theatre ticket, because the product he is buying is inherently hazardous to life and limb. Yet for air travellers, virtually all the factors that govern the safety of a trip are not available at the counter or elsewhere.

There is a common cause lying at the centre of innumerable accident investigations that will never be found among the sad debris of hardware at the site of an aircrash. It is the human desire not to tell, the instinctive urge to keep any blame-threatening facts under wraps. The motives for this are diverse but they lead to the same lethal consequences and they can be found in every area of the industry.

To start at the beginning, with the plane manufacturers themselves, one of the most important motives is the fear of corporate liability. The enormously high amounts of compensation awarded by the US courts to the dependants and relatives of aircrash victims are well known. The average rate of compensation for death is just under 1 million dollars per person (1988). Multi-million dollar accumulated damages no longer cause a gasp, and without mighty insurance cover airplane manufacturers and airlines would long ago have been pushed into bankruptcy. So it is not surprising that insurance premiums, hiked up to meet these inflated awards, are in themselves seen as a threat to the industry. For example, the leading corporations who supply business planes and light aircraft have already been brought to their knees, and some production lines have been halted. Meanwhile, the airline industry and its insurers which now face annual claims of over one billion US dollars (1988) are vocal members of the lobby campaigning for a reform of the present laws of compensation. It is a branch of jurisprudence that goes under the deceptively banal title of product liability. While international lawyers argue about the rights and wrongs of commercial liability, the practical effects of the present system have a sharp impact on air safety.

Airlines are no less jealous of their reputations than the plane manufacturers who supply their fleets, but pride is not the only commodity at risk. Carriers have learnt to be equally secretive in the face of ever-rising product liability claims and the crushing damages that can follow an accident. In the wake of the Zeebrugge Channel ferry disaster, the point was well made by an unnamed aviation lawyer to executives of a new airline who were complaining about the cost of meeting what they considered to be over-zealous safety requirements. 'If you think safety is expensive,' he warned, 'try an accident.'

Gagged

While the fear of facing such high compensation claims should act as a spur to higher safety standards, conversely it motivates attempts to cover up any facts that point to liability. Just how far this might have played a part in one of the most enigmatic and terrifying flights ever, has never been finally resolved.

Many American readers may remember the sensation it caused at the time, but its final analysis raised doubts that were less well advertised. It happened six years before Captain Ho's passengers were treated to their aerobatic display inside a Boeing 747 (pp. 19–21), this time to 85 travellers on a domestic flight aboard a smaller three-jet Boeing 727 who survived a very similar series of horrors after what must have been one of the closest calls on record.

Their Trans World Airlines flight from J.F. Kennedy airport, New York, on Wednesday, 4 April 1979, was scheduled for the 1,000-mile trip to Minneapolis, St Paul, but it never got there. Against all likely odds, the crew managed to make a successful emergency landing at Metropolitan airport, Detroit, Michigan – about halfway along their intended route. The B 727, one of the most ubiquitous types in the skies, has a solid safety record, but in the 43 minutes it was airborne it had been transformed into a battered and torn wreck with whole sections of its airframe missing.

What had happened – unknown to the pilots at the time – was that one of the eight leading-edge slats had extended from the forward edge of one wing while the plane was flying seven miles high at cruising speed. To put it briefly, this is an unthinkable event that perhaps shares the same order of horror as losing a road-wheel on a car at high speed. The slats or 'droops' that push out from the front of the wings, together with the flaps that extend from the rear of the wings, are there to provide the extra lift at slow landing at take-off speeds – but they should never be operated at higher speed. (On the other hand, the failure to deploy them during landing and, especially, take-off can be fatal. This happened to a Lufthansa B 747 in 1974 when it failed to lift off and crashed in flames on the edge of Nairobi airport, as we shall see in more detail later (pp. 118–27). For the moment, it is enough to outline the workings of the slat-and-flap mechanism.)

The forward slats are a clever device that generate a large amount of lift that, as it were, lifts the plane up by its own bootstraps. Without this effect heavy aircraft would not get off the ground in the allotted length of the runway. Essentially, they form a second layer to the curved front edge of the wing. When retracted, they fit unnoticed over the profile of the leading edge: when extended, they form a gap that ingeniously speeds the airflow over the prime lifting surface of the wing itself.

These forward slats are coupled with the more conventional trailing-edge flaps that can be seen sliding out behind the wing before take-off to increase the area of its lifting surface. The amount of the combined flap needed can be set in various degrees, according to weight, wind, length of runway and so on. The scary bit comes in the way these two systems are linked together. Both systems are controlled by one flap-setting lever. No one in his senses would attempt to extend the potent front slats at cruising speed. Both slats and flaps are designed for use at slow speeds only and to extend them at higher airspeeds over-loads their strength, eventually to breaking point. But a pilot might be tempted to extend the rear flaps a tiny bit to gain extra lift for a quicker climb to a higher level. How could he do so without the slats extending as well? He could pull an electrical circuit-breaker on the engineer's panel to disarm the slats, so that the flap lever could then only work the rear flaps.

Could this have been done on the TWA flight? This was an urgent question for the accident investigators, because if the pilots were not to blame, the possibility of a mechanical failure had worldwide safety implications for the 1,500 other B 727s then in airline service – then by far the world's best-selling jet airliner. Their attempts to reconstruct the sequence of events on the flight, however, met a curious snag. They had to rely mainly on the pilots' version of the flight, without the help of the cockpit voice recorder.

The pilots' account was that after some traffic delay at JFK, they had made a normal take-off into the evening darkness at about 8.30 p.m. Twenty minutes later they were pushing into a 100-m.p.h. headwind at 33,000 feet, and Captain Harvey Gibson got radio permission from ground control to climb above the worst of the wind to a calmer layer at 39,000 feet and so save some fuel. Significantly, perhaps, the plane was heavily loaded and needed a few minutes to burn off more fuel before it could make its climb, so meanwhile the crew continued eating their dinner undisturbed. Ten minutes later they started the climb into clear and smooth air.

Three more minutes passed before the trouble started. The onset came with a growing vibration that was soon followed by a steep bank to the right. Gibson disconnected the auto-pilot and took over the controls to check the roll. He couldn't. Despite his efforts, the nose of the plane dropped and the bank persisted until the plane was upside down and still turning and turning – through two complete barrel rolls. Just as the wings began to come level for the second time, the nose angle steepened and threw it into a violent spiral dive.

All hell was let loose in the passenger cabin. Free objects and drinks ricocheted about like projectiles among the screaming inmates, while the force of gravity pinned down their helpless limbs. Some were praying. Now the plane was halfway to earth in its seven-mile descent at a rate that exceeded all predicted stress and safety limits – but still the Boeing held together. Up – or perhaps one should say down – on the flight deck Gibson and his crew desperately tried every trick they knew to regain some control. They were flying through cloud at night, and so totally dependent on their instruments. But by now these were all far beyond maximums and were virtually useless; the hands of the altimeters spun round unreadably.

As the plane spiralled out of the base of the cloud layer at 8,000 feet, the sight of city lights whirled past the windows at unlikely angles. If the dive continued unchecked, Gibson and his passengers would hit the ground with the force of an aerial bomb in just fifteen seconds. Miraculously, he felt the control column beginning to respond.

What he could not know at the time was that the cause of his plight – an extended slat on the right wing – had suddenly been ripped off by the high-speed airstream of the dive. In a word, the massive forces that could have been expected to tear the plane apart during its gyrations had, ironically, finally served to save it from destruction.

Now the question was whether the wings would stay on as Gibson gingerly checked the high-speed dive. The Boeing is a tough animal and, amazingly, they did. Although many of the plane's systems had been badly damaged by the manoeuvres, he found enough control remained to attempt the emergency landing. When he drew it safely to a halt on the runway at Detroit, the waiting fire and rescue teams saw a shattered hulk of an aeroplane. Gaps showed in the fuselage, and the missing wing slat now declared itself to be the prime cause. But the mystery of how it had extended in flight remained to be resolved.

It never was – at least not beyond doubt. The main reason for this was that the cockpit voice recorder (CVR) – which should have registered all the talk between the crew for at least the last 30 minutes of the flight – inexplicably had been wiped clean. The accident investigators at the scene found that all it contained was a few minutes of talk recorded after the plane had been safely parked at the airport. According to rule, the nature of the flight was such that the content of the tape should have been preserved at all costs.

During the accident inquiry which followed, the captain explained to the National Transportation Safety Board that the erasure of the tape was a habitual routine carried out by most pilots after a flight. The Safety Board's final report, however, stated that 'we believe that the captain's erasure of the CVR is a factor we cannot ignore and cannot sanction': they had difficulty in accepting that the captain could have yielded to a habitual impulse 'after a flight in which disaster was only narrowly averted . . . our skepticism persists . . .'. The report went on to determine 'that the extension of the slats was the result of the flightcrew's manipulation of the flap/ slat controls'.

Yet the loss of the evidence from the CVR tape inevitably meant that the Safety Board's findings about the probable cause of the accident were short of ultimate proof: one member dissented from its conclusions about the erasure of the tape, but concurred that the crew had manipulated the slats and flaps.

In their attempts to re-create the events on the flightdeck, the investigators examine one putative theory. If the captain had decided to use a small amount of trailing-edge flap alone to assist the climb to 39,000 feet, he might have pulled out the slat circuit-breaker on the engineer's panel behind him, at a time when the engineer had momentarily left the flight deck. The flap/slat lever could then be safely used to feed on a little rear flap. When the engineer returned to his seat, he might have noticed the raised circuit-breaker and routinely pushed it back into place. This would command the front slats to extend (the NTSB established that four out of the eight had indeed done so, despite the force of the high speed airstream). At some time after the upset the crew, realizing the cause, might then have attempted to retract all the flaps and slats – but found that number 7 slat refused to retract, possibly because by that time it had been damaged by over-stress in the manoeuvres. The crew would then face the huge asymmetric pull caused by one isolated slat sticking out from the right wing – until it pulled itself off. The available evidence, however, could not positively establish this theory.

The US Air Line Pilots' Association expressed stronger doubts about the Safety Board's findings and stayed with the belief that a mechanical and not a human failure had caused the slats to extend, citing a large number of earlier B727 slat defect reports. One of these had indeed included slats extending at cruise altitude (this crew had retracted them before losing control).

If all the evidence from the flight had been made available, the NTSB, in common with most generous observers, would probably have given the benefit of the doubt to Gibson and his crew. In the last resort, they recovered from a desperate ordeal and landed their passengers safely. Only eight of them received minor injuries. But the absence of the recorded evidence leaves a question mark that is now never likely to be answered.

The evidence is more clear-cut when it comes to the fate of the Stone Canyon Band on New Year's Eve 1985.

The five-member band had performed for a couple of nights at Guntersville, Alabama, and boarded their plane there for the flight to Dallas, Texas, for the big night concert. They did not travel by airline but flew in a private Douglas DC-3 – an old 30-seat Dakota – owned by the entertainer Rick Nelson, who was on board with his fiancée. Their two pilots loaded the party and musical kit in the cabin of the vintage twin-engined plane and took off from Guntersville at midday. As they climbed out on an easterly heading, it grew chilly in the cabin and the pilots turned on the gasoline-fuelled cabin heater.

About an hour before the party were due to land at Dallas, the pilot radioed the Fort Worth air traffic control centre to say that he had smoke in the cockpit and wanted to divert to Texarkana, Texas. They heard no more from him, and a minute later the plane disappeared from their radar screens.

From then on, the story can only be told in the words of the two pilots – the sole survivors of the crash-landing that followed. Unfortunately, when interviewed separately, their two sworn testimonies were contradictory. The pilot said that he had gone back into the cabin and found nothing wrong with the heater when he checked it. After returning to the cockpit, the smoke got worse until it blacked out all forward vision. Through his opened cockpit window, he managed to select a spot for a crash-landing and put the plane down. When it came to a halt he escaped through his side-window. There was, he said, a small fire inside the cabin but when he shouted to the passengers there was no response, so he started to search for them outside.

The co-pilot's version of the sequence of events was much more descriptive. When the cabin heater 'acted up', he said that the routine was to turn it off, wait a while, and then sneak it on again. The captain had told him to do this cycle several times until finally he got nervous and refused. Thereupon the captain left his seat again to tinker with the heater and signalled back to the cockpit to turn the fuel on once more – and this time his mate obeyed. Soon after, passengers came forward to the crew complaining of the smoke in the cabin before it finally enveloped the cockpit. After the crash, the second pilot recalled clambering out of his side-window and falling to the ground. Staggered round to the cabin, he saw a blazing inferno inside it and, fearing an explosion, he ran clear.

He then bumped into the captain, who sat him down on the ground and entreated: 'Don't tell anyone about the heater, don't tell anyone about the heater. . . .'

All of the seven people in the cabin perished in the flames. Faced with the conflicting evidence, the Safety Board found that the source of the fire could not be proved, but blamed the captain for failing to apply the approved fire drill. This might not have prevented the loss of the plane, it said, but it would have increased the chances of survival for the passengers. For our purposes, any moral that can be drawn from the human frailties shown on that day are obvious enough, and need not oust a touch of compassion for the two men who live with the memory of the disaster.

Of the 10,000 Dakotas built, some 800 are said to be still in service. If the captain's plea to keep quiet about the heater malfunction had been answered, other crews would not have been alerted to the danger and it might have happened again. In the event, the co-pilot disobeyed his captain and came clean. Individuals, it seems, are less tolerant of secrecy than corporations.

6 | The law at work

The potentially vast claims for compensation that can follow an aircrash are not the only reason why the laws of product liability can put the facts out of reach to those concerned with air safety. It needs a glance inside the legal undergrowth to see more clearly how this has come about.

Leaving aside much detail, there are two conflicting principles of law at work. The first declares that anyone whose business it is to put a dangerous thing 'into the stream of commerce' remains absolutely or strictly liable for any damage that ensues, irrespective of whether or not he is proved to have been at fault. This is the principle of 'no-fault liability'. In other words, the law seeks out the agency seen to be *most responsible* for creating the risk in the first place (and which benefits from the profits of the enterprise).

The second and converse precept is that such a person shall be liable only in so far as he can be proved to have committed a fault: this is usually a brand of negligence or wilful misconduct. It clearly puts a much heavier burden on claimants and their lawyers, who need access to evidence – mostly possessed by the commercial defendants – if they are to have a chance of proving blame.

By and large, the no-fault rule operates in the USA (apart from variations in some state laws), and since most airliners are made there, a claimant's best option is to sue the US manufacturer or airline in those courts. In Britain and parts of Europe where the second and narrower law prevails, they are faced by the need to prove blame. The legal costs of proving negligence or fault in the English High Court can run to £10,000 a day, and although the scale of damages awarded there is becoming more generous, it remains far lower than it is in the US courts.

Consider the unusually bold British air passenger who sued his airline after breaking his spine when the plane cruised through a patch of turbulent air on a flight to Bangkok. He needed to prove that the pilot or the airline had acted recklessly. After a hearing that ran into weeks, the High Court in London accepted his proof and awarded substantial damages and costs. The airline's insurers, with a deeper pocket than the crippled claimant, took the judgment to the appeal court and reversed the first decision, leaving the bankrupt victim facing combined costs of over £50,000. (The defendants graciously waived a good part of their claim in the event.) Under the no-fault rule, it would have been an open-and-shut case in favour of the claimant: proof of the cause of the injury would have been enough, whether or not anybody was to blame.

The result is that in Britain and other jurisdictions where proof of fault is essential, few people can contemplate the financial risk of going to court (public legal aid is available in the UK to the least wealthy, but the limit is low enough to exclude most people who use air travel). The few who do go to law tend to be those who have enough cash and temerity to risk it, or the less wealthy or State-subsidized litigants whose lawyers are confident of winning on a strong case.

Prudently, the plaintiff's lawyers will do all they can to settle out of court and save the vast costs of a hearing – a recent inquiry showed, for example, that in the UK only 10 per cent of tortious claims ever reached the courts. So that in most cases the facts about an aircrash and the identity of those responsible for it – however vital that evidence might have been for the future safety of other air travellers –are never made public. The defending company or airline thus not only avoids adverse publicity but pays out less than might have been awarded against it in court, as well as saving the costs of an action. Any benefit for the claimant is limited to an immediate cash handout under the private settlement, which will almost certainly be less than a court would award if the case had been won there.

Instant cash in hand is an attractive prospect. If the claimant risked a High Court action, they would be lucky to get an answer within five years. But the option of an out-of-court settlement conceals another hidden snag that is perhaps less damaging to the claimant than it is to the rest of us. Lawyers defending the reputation of their clients have a more sinister trick up their sleeves that, although ethically under question, is becoming more common. They insert a total secrecy clause as part of the settlement deal that binds the claimant to silence, not only as to the amount paid to him, but also covering all the facts revealed in the case papers – evidence, admissions, discoveries. It means, of course, that information that could have a vital bearing on safety and crash-prevention is forever put out of reach. Although there is a good deal of unease about the propriety of such clauses – notably in current drug injury cases – it has not yet inhibited the spread of such lip-sealing compacts.

Secret settlements happen, of course, on either side of the Atlantic. What about the merits of the two legal systems – no-fault in the US and proof of blame on the English pattern? On the face of it, most people would probably think it unjust that companies or individuals should be made responsible for things that are not their fault – besides being commercially damaging.

Exponents of the no-fault rule can point to a whole lot of established law based on the principle of strict liability for dangerous activities – industrial injuries, for example. It has a respectable pedigree and the seeds of the idea existed in Roman law before it became written into the Code Civil of Napoleon's France which, being irreverently translated, simply said that you are automatically responsible for any injury or damage to others caused by your hazardous activities or belongings. In England, a part of this concept has long gone under the simple and charming title of liability for 'dangerous things', long before airplanes might have been added to the list of ferocious bulls, rabid dogs and collapsing masonry. But British laws have kept civil aviation off the list so that victims or their dependants still must prove blame to win compensation.

British reluctance to extend the no-fault principle ensured that a recent European move towards it was still-born. After years of study among international lawyers, the EEC produced a draft Directive that would have given its citizen-members the right to compensation simply on proof of damage alone, as in the USA. In Britain, industry won the argument against it largely by invoking fears about the crippling penalties faced by its American counterparts. Backers of the EEC reforms pointed out in vain that these were not due to the no-fault rule itself, but to three other features of US legal procedures. Awards are much higher there because they are made by juries and not judges; punitive damages can and usually are added to inflate the total (uncommon elsewhere); and finally, the USA has a double-or-quits system of lawyers' contingency fees. Unlike most European countries, US lawyers commonly charge 20 to 35 per cent of their client's award if he wins, and nothing if he loses. Critics of the system see this as not only an encouragement to the ambulance-chasing type of litigation, but as a built-in and improper incentive for lawyers to win unconscionably high awards from the jury, with a juicy percentage as their prize. Others see the benefits of a self-funding system by which loser-clients pay no lawyer's fee.

Shorn of these overtones which are special to the USA, the no-fault rule has been shown to work fairly elsewhere. The resulting increase in insurance premiums – constantly prophesied by industry – has been almost negligible. West Germany has applied it to its pharmaceutical industry for ten years without unduly inflating the level of damages, and Sweden has adopted a similiar approach. The most persuasive example comes from New Zealand, which has applied the rule across the board under a statutory insurance scheme: all those shown to have been injured or damaged by an accident of any kind are compensated at fixed rates without further ado in the courts, but if manufacturers or others are found blameworthy, they can be ordered to pay a contribution to the public insurance fund.

To put it politely, Britain rejected the EEC proposal recently because of a misconception. The government was persuaded that it would unfairly damage industry to expose it to penalties without proven fault on its part, and it relied on fears from the American precedent. It had not read the US scene in context, nor did its industrial advisers want to hear too much about the success of no-fault systems outside America.

Nor did they recognize that the legal argument for the no-fault rule had been fought and won. A string of royal commissions and other learned statutory bodies had long since reached a consensus in its favour, as had the confederacy of EEC and international lawyers.

Accident liability is a complex mixture of law and when its many tributaries are poured into the international melting-pot it must appear as an inscrutable brew to laymen. Whatever their advisers thought, it is little wonder that government ministers simply did not understand the issues when they came to debate in parliament. And to be less than handsome about it, many more votes lay with the big battalions of industry than among a few potential accident victims, still fewer of whom would contemplate going to law in Britain under the existing costly system that demands proof of fault. It is not a benign proposition for air travellers faced with a hugely technical and well-heeled adversary.

There is another ugly strand of international law that makes the compensation of aircrash claimants even more arbitrary and capricious. Long ago when fledgling airlines first emerged with small or shaky resources, there was a recognized need for a guaranteed amount of accident compensation for international air travellers. The Warsaw Convention, followed by other treaties, required airlines to insure to these limits. It is ironic that as time passed, instead of ensuring that the public got fair compensation, the international limits have now come to protect airlines from the much higher – and more realistic – damages that the courts would otherwise award. Subject to a number of circumstances, the international limit is set as low as £13,633 for air passenger injuries or losses of a kind that – in one recent road accident – brought a British claimant a court award of £850,000.

The Warsaw and other international conventions now represent a lamentably low threshold for fair compensation, but there are at least two loopholes. First, the limits do not prevent higher claims if an accident is proved to have been due to the intentional or reckless misconduct of the carrier. (The Bangkok passenger who had his spine broken managed to establish this in the lower court, but it was rejected on appeal – such are the dubious benefits of litigation in English courts.) Second, some countries and some airline routes – notably US domestic routes – are not bound by the conventions. Again some countries and some airlines have adopted 'special contracts' which increase the basic Warsaw limit from £13,633 to as much as £50,000 (British Airways, for example). So it can be seen that air accident claims are a lawyer's paradise – the outcome can depend on which countries are involved, the airline and the type of route.

The most that can be said is that a claimant's best hope is to have flown on an internal US domestic route – thus avoiding Warsaw limits – and on a US airplane and airline, so that his case will be accepted in those courts under the more liberal no-fault law. The rest of us just need a good lawyer and a thick enough bankroll to hire him.

Safety needs constant public vigilance. And public awareness – the mainspring of effective action – depends on public information. Outside the USA, the legal systems ensure that it is in short supply.

There is not a happy prospect for the best standards of safety in the air as long as the industry is so well protected from the financial sanction of having to pay up for its mistakes and mishaps – however they may happen. The burden of the no-fault law is that it is the industry's choice of earning a living and airplanes still prove themselves to be dangerous things. It may appear to be a brutal view, but death and injury are equally brutal.

7 | International cover-ups

It is not surprising international law is so muddled. Whether or not the cosmic anxieties about the future of the planet may come to speed the pace of international co-operation, for our time at least it seems that the business of foreign affairs will remain chiefly as an inter-state contest in chauvinism. No diplomat allows his country to lose public face if there are ways to avert it, whether this is achieved with rhetoric, secrecy or worse.

Like the United Nations, the International Civil Aviation Organization (ICAO) as a Special Agency of the UN plays the role of umpire within the known limitations of its parent organization. As always, its effectiveness must rely on the consent of all its member states, and this inevitably spells compromise and delay: the laggard ship sets the pace of the convoy. ICAO, for instance, does lay down the ground rules for accident investigations as well as many other agreed safety standards. But they are minimums.

Foreign languages

Take, for example, the use of English as the language of the air. There is an obvious need for a universal language and, largely by an accident of history, it was English that first led the field. But the resentment about this *fait accompli* has never been quenched and flares up from time to time. ICAO had to accept the inevitable – up to a point. While its members agreed that English should be the international language, a concession was added which allowed nationals to use the local language in their own airspace. The uneasy compromise now remains by which any pilot may request English to be used and this must be met by the air traffic service of all member states. But this does not apply to other *pilots*. The result, of course, can be both chaotic and dangerous. There are plenty of examples, but here is a recent complaint from a British pilot.

89

When there are delays before landing, pilots are often told to enter the hold or 'stack' over a radio beacon or a similar position fix. They then circle around a race-track pattern at given heights – 1,000 feet between each is the minimum safety separation. Each successive plane in the landing queue is then picked out by the controllers, much like selecting plates from a kitchen rack. The British pilot in this particular stack does not tell us where he was, but the nationality of the airport is clear enough.

He was told to enter the holding pattern at 6,000 feet, with a DC-9 already circling below at 5,000. After one circuit, another DC-9 entered the hold above him at 7,000. Wedged closely between the two other airliners, he tells how for the next five to ten minutes there was continual chatter in Spanish between both the DC-9s and the Spanish controller. Messages on the same radio channel to other planes landing, taking off and taxiing at the busy airport added to the confusion. The captain complains that he found it impossible to establish any sort of mental picture of the movements of which he and his passengers were a living part ('blind' flying by instruments – nearly always the case – demands a constant appreciation of everyone's relative positions and intentions). He concludes that the enforced use of English alone would boost flight safety 'and very possibly avoid a future accident'.

It is an old, old story. In the same month – August 1986 – several airline captains complained that it is a tendency that is growing in Spanish airspace, so much so that one of them suspects that it has become 'policy' to spurn English. Non-Spanish crews are unable to monitor or understand events, to query an error or check another pilot who reads back an order incorrectly and, he says, 'We all know the potential hazards of this.' A third pilot reported that twice in one week British and German planes were enveloped by a 'tirade of Spanish – we had no idea what was being said'.

Such is the real level of international co-operation in the air. One might have hoped that the urge for survival in such a demanding realm – if in no other – might have put sense before chauvinism. Apparently that day is not coming any closer.

Spain is by no means the only country that allows political self-esteem to infect its airspace. The French are nothing if not great patriots. They are also doughty strikers, as continental air passengers know from delays caused by industrial troubles that afflict their air traffic controllers from time to time.

Communication breakdown

One such occasion occurred in February 1973 at the height of the Spanish sunshine tourist season in Europe. French air traffic controllers, as state officials, were then prohibited from taking strike action and their campaign to win this right had so far failed. Now chaos in the air was threatened as they defied authority by striking at the busiest season. The government, seeking both to neutralize the industrial clout of the strikers and to keep the tourist traffic flowing, ordered the existing military radar and air traffic control network to handle all civilian flights over France under a diktat of its transport minister, M. Robert Galley (see p. 96).

Most airlines with routes into France or through its airspace to the Spanish beaches asserted that the handover to the military network was no threat to safety. Pilots' associations expressed consternation but continued to fly, although some of the more cautious airlines reacted by re-routing flights west of the French coast over the Bay of Biscay. The French government's assurances that the military could cope with all the traffic was rather naturally denied by the usurped civilian strikers, who took it as further provocation. Having thrown down the gauntlet, the political stakes had multiplied for M. Galley and his officials with the need to show that the skies were indeed safe in the hands of the military.

At lunchtime on Monday 5 March two airliners collided in mid-air over the city of Nantes on the west coast of France. One, an Iberian Airlines DC-9 bound for London from the Spanish island of Majorca, crashed and killed all 68 people on board. The other, a Spantax Convair Coronado with 99 passengers on board, lost a large chunk of wing. Although severely crippled, it miraculously managed to make a safe emergency landing at the nearby military airbase of Cognac. Later when the full story became known, it was held to have been a most skilful feat by its pilot, Captain Arenas, but when he walked from the stricken plane he was immediately put into detention by the military commandant of the airbase.

World attention had focused on the crash, and the French authorities, now pushed into a corner, chose to defend the competence of their air force controllers seemingly at any cost. The Quai d'Orsay in Paris insisted that the collision had been due to 'a succession of pilot errors' by Captain Arenas and that the military controllers had been 'completely exonerated' from any blame for the mid-air crash. This, it was claimed, was proved by an examination of the tape-recorded exchanges between the pilots and the air force ground controllers, but the text of these remained a military secret and were not released to the press.

That week an investigation published by the *Sunday Times* in London claimed that most of the blame for the disaster fell on the French military controllers and the dangerously inadequate system they operated. They had realized – far too late – that the planes were converging. The report showed that the collision occurred because the controllers then gave an impossible order to Captain Arenas and also left the DC-9 pilot completely unaware of the danger he was in as the two planes converged. Amongst other faults in the control system that made the disaster a virtual certainty was that at the crucial time the two pilots were instructed to listen to separate controllers on different radio frequencies. Neither of them could therefore know that a mistake had been made. Under the civilian regime both pilots would have come under a single controller at Bordeaux, so that all three men would have been kept in the picture.

Under the military set-up, although the airspace around Nantes was covered by at least one and possibly three radar ground stations, neither pilot was warned at any time that he was heading for disaster. Nor did the military observe the civilian rules that ensure a safe separation between aircraft: that is, either 1,000 feet between them vertically or ten minutes' flying time if they are flying at the same altitude. In fact, during the lead-up to the collision, the time-gap between the Coronado and the DC-9 was only four minutes and both pilots had been allotted the same level at 29,000 feet.

Some months after the crash, the London *Times* reported that the French air force was trying to suppress the findings of an inquiry chaired by the chief civil aviation inspector which had concluded that elementary errors by the controllers had caused the crash. A draft of the findings had been leaked to civilian officials, who believed it would never be made public. And when the accident report was published two years later, it indeed turned out to be a filleted version, shorn of a full transcript of the crucial radio messages that are normally set out in full in such documents. An outline of the fatal flights shows there is a good deal in the claim that the collision was actually caused by the controllers.

The doomed Iberia DC-9 had left Palma in Majorca, the Spanish Mediterranean island, at 11.24 in the morning bound for London. Cruising at FL 310, it would cross the Pyrenees to join the flow of holiday traffic over western France. After entering French airspace, at 12.32 the pilot was told to descend to FL 290 and he reported back his estimated time of arrival overhead the Nantes radio beacon as 12.52. Due to varying winds, at 12.41 he quite normally gave the controllers a revised estimate for Nantes of 12.54. The collision happened at 12.52.

Meanwhile Captain Arenas in his Spantax Coronado had left Madrid at 12.01, also for London. By about 12.20 he too was crossing the Pyrenees cruising at FL 260 and after contacting the French controllers, at 12.30 they told him to climb to FL 290 – the same level as the DC-9 which, unknown to him, was converging onto his track, as it were, over his right shoulder.

At about this time there were some five aircraft on the same route, some using one radio channel and some another, some speaking in English and some in French. (Only the captain of a British Vanguard seemed to be speaking his native tongue, and he anxiously tried to call his controller five times, through a haze of chatter, to confirm the flight level at which he was supposed to be cruising.)

Arenas was also having difficulty in getting things clear. At 12.43 he thought he had heard an instruction *not* to arrive over Nantes before 13.00, while his present estimate for the beacon was 12.54 – the same as the DC-9. He did not know that this was the reason for controllers' attempts to slow him up. It was an impossible command. Now with only some ten minutes to run, he could not possibly lose six minutes by decreasing speed. At high cruising level, a laden airliner can only cut its speed marginally without stalling and falling out of the sky. If the order to delay was correct, he would have to do something drastic.

Now with the help of hindsight taken from the pages of the Spanish transcript of the air-to-ground radio exchanges made on that day – still bearing the stamp 'SECRETO – CONFIDENTIAL' – the countdown to the crash continues.

Arenas calls: 'Please confirm we must report Nantes at 13.00.' But

93

he can get no reply during the next five minutes, while no less than 25 other calls from and to other aircraft blank out his radio channel – including the plaintive cries from the British Vanguard.

At 12.49 – with only five minutes left before crossing Nantes – he is asked to change radio channels. When he gets through to the new controller, he is *in extremis*:

'We request to make a three sixty [full 360-degree turn], a three sixty to the right in order to pass Nantes at one o'clock, one three zero zero (13.00) . . .'

No reply. Marina is the French controller's call-sign, his flight number is BX 400. At 12.50 he calls again:

'Marina? BX 400.' And again:

'Marina? BX 400.'

Arenas does the right thing: he follows the last instruction given to him, albeit heard indistinctly. He must assume from it that other traffic will be funnelling into the radio beacon and it will not be safe to cross it before 13.00. So just before 12.51 – with only three minutes left – he takes the only option left. Again correctly, he signifies his intention 'blind' to Marina control, adding his flight level as a further warning. It is now 12.51. Bad visibility adds to the tension.

'For this delay we're turning to the right a three six zero degrees maintaining two nine zero (FL 290).'

Arenas again gets no response while the controllers talk to two other planes. (He had started to orbit to the right because he knew that the busiest air corridor from the Canary Islands was converging on Nantes from his left. The French later tried to fix blame on Arenas by arguing that the ICAO rule is to orbit to the left, unless advised to the contrary: had he done so, they claimed that the collision would not have happened. Others denied this rule. In fact, the ICAO code is silent or at least ambiguous on this point. At any rate, as the Coronado raced into danger with only seconds to spare, no pilot could be expected to solve a legal riddle that beset the post-mortem experts themselves.) Arenas swung his plane into a right-hand orbit.

Just before 12.52, and expecting Marina control to reply, he thought he heard his call-sign through the radio chatter (the record only shows an exchange with IB 163 – another Spanish plane). Hopefully, Arenas snatches at this straw:

'OK. Marina, (BX) 400 – go ahead?' Then, getting no reply, he makes his last recorded call just after 12.52:

'Go ahead Marina, go ahead, this is BX 400, go ahead . . .'

But there is silence. Seconds before the collision with the DC-9, all he hears is another British charter plane trying to raise Marina:

'Marina Radar, this is Court Line Golf-Alpha-Xray-Mike-Hotel. Good afternoon?'

There is no reply from Marina to either plane before the DC-9 exploded in mid-air as it flew head-on into the left wing of the turning Coronado – at 12.54. Even when Arenas transmitted a Mayday call at 12.56 from his crippled plane, it was not heard by the controllers but picked up by the pilot of IB 163 – the other Spanish charter flight – who later managed to relay the alarm to the ground.

Amongst its other strictures, the French transport ministry report of the accident blamed Arenas for making his turn without authorization from Marina control.

There is a further sidelight on the morning's work of the military controllers. Ground control centres are linked by telephone land-lines to hand over control of each plane as it moves through successive regions. Whatever reliance should be put on the uninhibited pages of the French satirical weekly *Le Canard Enchainé*, the conversation between two air force controllers who were directing the traffic suggests a remarkably light-hearted attitude to the task in hand. Although I cannot vouch for the text it prints, its source is well known to me and comes from one at the centre of the affair whose claim to have seen the records in question need not be doubted.

Le Canard claims to give the tape-recorded exchange of the hand-over from the southern centre at Mont-de-Marsan to the region to the north controlled from Brest. Seemingly unaware of the collision that had occurred 22 minutes earlier, they are both trying to trace the Coronado:

Mont-de-Marsan (M):	'You've had no reply?'
Brest (B):	'No.'
M:	'Good, understood . . . continue to call him.'
B:	'Oh well, as if we haven't got enough on our plate as it is!'

M:	'Well all right, but you can't just leave a plane like that!'
B:	'But we've never seen it – you must be joking!'
M:	'No, I'm not joking, no!'

Le Canard's comment on this exiguous exchange is that there was indeed nothing to joke about. And it is worth remembering that the collision at Nantes had not come unheralded. At the time French airline pilots, who in the six days before the crash had reported no less than 20 air misses, shared the *Sunday Times*'s findings. The British Airline Pilots' Association (BALPA) imposed a ban on flights over France immediately after the crash was announced. British Airways (then BEA) bowed to the inevitable and cancelled its French routes, soon to be followed by the Dutch airline KLM.

The official French report of the accident, published a year later, admitted that the military control system had 'created a source of conflict' and mentioned 'difficulties' in radio communications which resulted in 'a complete failure of the crews and control to understand one another'. Beyond that, its authors were content to shy off from the unavoidable conclusions, much less to put the blame where the facts located it. Meanwhile, the French press had accused the transport minister of lying, and the aviation world drew its own conclusions. And it is fair to say that after a decent interval of two months, long enough to wipe the immediate memory of Nantes off the political scene, M. Robert Galley ceased to serve as France's minister of transport. It was even longer before the relatives of those killed in the DC-9 finally won compensation from the French government. Some five years of legal wrangling ensured that the affair had faded safely into oblivion before the court settlement tacitly acknowledged the responsibility of the military controllers.

Another much graver chapter was to follow that struck a distant echo with the Nantes disaster. Two years after the sinking of the Greenpeace ship *Rainbow Warrior* by the French secret service, an international arbitration tribunal awarded £5 million for the one death and damage caused by the clandestine attack. It revived the events of July 1985 when the Greenpeace ship was lying up in the French Polynesian island of Mururoa and, while it prepared for demonstrations against underground nuclear tests, it was bombed

96

and sunk during the night. France protested its innocence and, like any other state, disowned its agents until evidence of their complicity forced a belated admission for what was then seen to be a deliberate state crime.

It is perhaps understandable that the French posture of injured innocence only admitted one error – that of being found out. That ethic has a long ancestry stretching far beyond the Gary Powers U2 flight over Soviet territory in 1960 and forwards to inspire many more contemporary intrigues. There is not much left to surprise us in that line of business, except perhaps an ever-diminishing faith in State rhetoric. The shock comes when the black arts of foreign policy are seen to pervert the cause of international air safety: when the life and limb of all – regardless of nationality – so clearly depend on access to truth unhindered by any misplaced patriotism. It is a thought that may suggest an epitaph, not only for the 68 passengers killed at Nantes, but equally for 144 Britons who died in the mountains of Tenerife in the Spanish Canary Islands eight years later.

Confusion at Tenerife

The search for the true cause of that disaster led to an extraordinarily bitter dispute between Spanish and British investigators that only came to light through a bizarre event that lies on the borderline between legitimate newsgathering and the more restricted territory of diplomatic confidences. Besides a partisan slanging-match, the story has an added twist in the tail that reveals much about the British love for official secrecy. It began to unfold as soon as the first news of the crash came over the wires after lunch on Friday 25 April 1980.

The Dan-Air Boeing 727 charter flight from Manchester, England, had been uneventful until it began its approach to Tenerife North airport (then called Los Rodeos) in the Spanish Canary Islands. Its air traffic controller told the pilot to descend to 5,000 feet and expect a routine landing without delay. The pilot obeyed, but as the plane flew down below the height of the 7,000-foot mountains lying close to the west of the airport, the controller changed his mind. He had realized that the Boeing was now overhauling a slower Iberian Airlines Fokker 27 which he had placed first in the landing sequence.

What he should have done and what he did do were, unfortunately, two different things. He *intended* to tell the Boeing to take up a holding pattern over a radio beacon sited on the airport. (See Figure 1.) These 'holds' can be flown either as right- or left-handed circuits and the direction is shown on the pilot's printed landing chart. But, for reasons that will be seen later, there was no chart. So it was up to the controller to guide the pilot to take up either a left- or right-hand circuit. He *should* have radioed 'the standard holding pattern turns to the left' or more briefly 'left-hand pattern in the hold'. Instead, as later recordings revealed, among the words he used were 'turn to the left'. The crew took this as an instruction to turn left forthwith as a preliminary before entering into the hold over the beacon. But that ambiguous and fatal instruction to turn took the Boeing towards the mountains.

It may be hard to understand how an airliner can be put into such total jeopardy by a few misplaced words, but a closer look at everyday techniques may help to explain it. The pressures of time and traffic call for a universal phraseology that almost amounts to a code. This is laid down by ICAO. In theory it must be rigidly observed by pilots and controllers – and in the best circles this is done. A slight variation – even in the order of the words used – can therefore spell a gross misunderstanding.

For example, more often than not there are several holding patterns near airports. As we know, aircraft fly their race-track course with a radio beacon or position fix exactly marking one corner of it – the whole circuit taking four minutes to fly. There are a number of other formal procedures that need not bother us here, but all are printed and distributed to airline pilots and controllers. One of the objects of placing holds and procedures so precisely is to keep traffic safely clear of high ground so that each has a minimum safety altitude allotted to it, and indeed all other airspace sectors and airline routes must also have this vital data marked on the charts. An approach chart for Tenerife North, for instance shows a minimum of 6,600 feet for the inwards route over the sea from the north-west and 14,500 for the mountainous sector to the south-west. The chart marks one of the approved holding patterns safely placed over the sea to the east of the airport showing a minimum altitude of 5,000, despite the fact that there is high ground rising to 4,950 only five miles to the south-east of the airport. The margin for error may seem

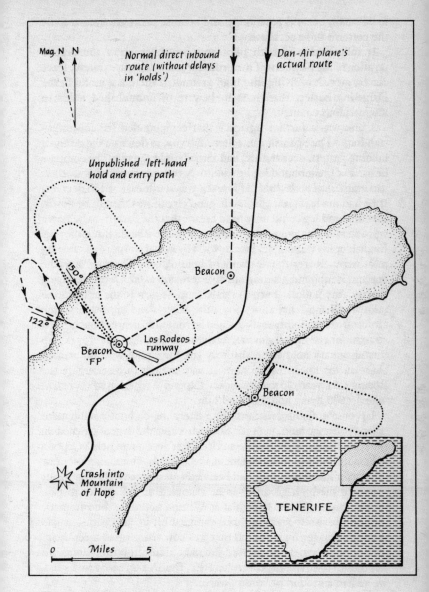

Mag. N   N

Normal direct inbound
route (without delays
in 'holds')

Dan-Air plane's
actual route

Unpublished 'left-hand'
hold and entry path

150°

122°

Beacon

Beacon
'FP'

Los Rodeos
runway

Beacon

Crash into
Mountain
of Hope

TENERIFE

0   Miles   5

Figure 1. Diagrammatic plan of Dan-Air flight on 25 April 1980

to be small, but it is acceptably safe because pilots are trained to fly the patterns most accurately.

It follows that if both pilot and controller know the precise position of the plane and if the procedures are flown accurately, there can be no risk of flying into high ground. If there is a moral in the Dan-Air disaster, then it is to beware of unpublished holds in mountainous country.

There was a further intricacy that compounded the misunder-standings. The Spanish controller, in trying to describe the notional holding pattern over the air, had initially told the pilot the compass bearing of the inbound leg he should fly towards the beacon *when he was established in the hold*. This track was south-east or 150 degrees. If he had made it clear that a left-hand circuit was then to be flown, that inbound leg could only have meant that the pattern to be flown was safely out to sea to the north-east of the island. But either his English or his training let him down. His actual words to the Boeing pilot were: 'Roger the er standard holding over Foxtrot Papa (the beacon) is inbound heading one five zero *turn to the left* call you back shortly' (my italics). Turn to the left means turn to the left and the pilot – still some five miles from the beacon and approaching the airport from the south-east – turned left onto 150 degrees and headed straight for the mountains. Defective as it was, the order to take up this imaginary hold came absurdly late for a pilot with no printed chart on his knee to help him. It was enough to disorientate the Boeing's 51-year-old commander, Captain Arthur 'Red' Whelan, with 15,000 flying hours behind him.

Up on the flight deck a minute later, the co-pilot commented: 'Bloody strange hold, isn't it?' Soon after this the danger dawned on the crew and the flight engineer cried: 'Let's get out of here!' Captain Whelan, sitting in front of him at the controls, simultaneously saw the horror into which they had been led and shouted: 'He's taking us round to the high ground!' as he slammed on full power to gain height. But it was too late. Just as Whelan said this, the automatic ground proximity warning horn sounded off in the cabin – it had responded to the high ground that was now rising up to meet them. With a reflex action, he pressed the radio transmit button under his left thumb to advise the controller: 'Er, Dan-Air on zero zero eight – we've had a ground proximity warning . . .'

They were Whelan's last words. Four seconds later the plane

struck a 5,450-foot ridge below the 7,000 summit known as the Mountain of Hope, killing all 146 people on board. Eighteen months later, the Spanish official report of the accident blamed Captain Whelan. The cause, it said, was that he flew into high terrain regardless of his altitude.

In truth, Whelan had been bidden to do an impractical, indeed impossible, manoeuvre as a result of the controller's late change of mind about the landing sequence. The Spanish authorities had not published any standard chart of the holding pattern. That left it up to the controller to describe the procedure precisely and in good time. He did neither.

The airline Dan-Air brought legal proceedings against the Spanish controllers, and the British authorities took the rare step of publishing an addendum to the Spanish accident report. This declared, in language moderated by diplomatic convention, that the instructions given to Whelan were 'ambiguous and contributed directly to the disorientation of the crew'. That phrase concealed an exceptionally raw behind-the-scenes wrangle between the British and Spanish governments in the months leading up to the publication of the report. Had it not been for this determined British intervention, the Spanish official version of the facts might have been as baldly nationalistic as M. Robert Galley's exoneration of his French air force controllers for their part in the collision over Nantes seven years before.

The British addendum to the report – in fact written by inspectors of the Accident Investigation Bureau (AIB) at Farnborough, England* – included another devastating criticism of Spanish safety standards. They pointed out that even under the basic ICAO code, the minimum safety altitude for the entry manoeuvre was 7,000 feet, and for the hold itself 6,000 feet: and that a further 1,000 feet should have been added to both minimums as recommended in hilly terrain. Whelan should not have been cleared below 8,000 feet until he was safely into the hold at not less than 7,000 feet. Had the Spanish controllers not told Whelan to descend to 5,000 feet, they said, 'this accident would not have occurred.'

They went on to say that as there was no published landing chart, it was reasonable for Whelan to assume that the crucial calculations

* Since re-named the Air Accidents Investigation Branch (AAIB).

for the lowest safe altitudes had been made and applied. But no evidence came to light that 'this had been carried out by a competent authority'. The Spanish had also blamed Whelan for not flying the 'ideal' track which, with defensive hindsight, they depicted in their report. The British inspectors dismissed this by showing that it was impossible for an aircraft to 'fly around the sharp angles drawn'.

If the substance of these criticisms had been included in the Spanish version of the accident report, the AAIB remark that 'it would have been acceptable to the United Kingdom'. Otherwise not.

Soon after the accident in April the *Sunday Times* Insight team published a reconstruction of the flight: we had grasped the implications of the missing landing chart and the misunderstandings between pilot and controller that led the plane onto its fatal course towards the mountains. We deduced that Whelan had been asked, at a disconcertingly late stage, to fly a pattern from a chart that was not to be found in airline flight manuals. Later it turned out to be an entirely informal manoeuvre that had not been officially approved or published. But the full proof of this version of events lay in the transcript of the recorded exchanges between pilot and controller and this still remained under lock and key. Not for long.

Three days after our story was published, I returned to the office after lunch to find an anonymous buff envelope stuck into the keys of a typewriter. Inside it there was an innocent-looking tape cassette which I thought might be some publicity spiel or perhaps some office prank. But when I played the tape I was stunned to hear the instantly recognizable voice of the chief inspector of aircraft accidents at the AAIB. But his tone was much less familiar. It was one of outrage, exasperation and tenacity – while his foreign adversary at the other end of a telephone line was defensive and clearly getting the worst of a long verbal skirmish. The subject was immediately evident.

Those who have not worked in newspaper offices will probably remain forever sceptical about such unsolicited windfalls which, according to popular image, do not fall from the skies without some complicity by journalists. In fact anonymous leaks are common enough but their value is clearly limited unless they can be identified or checked out. But in this case the source guaranteed itself, although the hand of the intermediary remained unknown. To this day I have no idea who the donor of the cassette was, nor his or her possible motive. But I have a feeling that it was a benign one. I am, however, as certain as can be that it did not emanate from the AAIB itself. Its reaction to the leak proved as much.

The tape of the telephone conversation confirmed the backbone of our deductions and revealed much of the actual detail of the flight, quoting blow-by-blow exchanges between the Boeing's crew and the controller that had led to the frightful misunderstandings. The chief inspector admitted that Whelan could be said to have made navigational errors -- the pilot is always ultimately responsible for his own navigation and for knowing his exact position -- but he refuted the Spanish insistence that the hold he was asked to fly had been approved: it had not, and there was no published chart as there should have been. In fact, said the chief, the pilot had not only been given the wrong holding pattern but one that was impossible to execute: the order had come too late, and in any event a Boeing could not comply with the tight turns it demanded. On the question of safety levels, he crashed into the Spaniard for the breach of this cardinal regulation. The chief followed up by forcefully rejecting the Spanish refusal to accept any wrongdoing on the part of their controller.

This fraught telephone discussion came at a time when pressure was building on government ministers in the House of Commons to make a statement about the progress of the inquiry. Protocol demanded that both the governments concerned agree a common text not only for the accident report but also for any public statements on the affair. So a hiatus persisted while the Spanish remained adamant that the British pilot was to blame. Despite these political pressures, the chief stuck to his guns and it was clear that he was not prepared to allow the UK to put its name to a partisan and distorted version of the accident. As the voice from Madrid remained unyielding, the British voice became more insistent:

'If we can't get agreement, I will give my minister what I believe to be an accurate thing and he will say it -- and to Hell with the consequences. Do you understand? We will do it by consulting the Foreign Office because there will be a row, won't there?'

He went on: 'The trouble is that my minister cannot have a situation where 146 people have been killed and there is no statement to the House. In fact, I'm surprised the newspapers have given us as much peace as they have, aren't you?' The Spanish voice agreed. The chief explained how the parliamentary process in Britain does not easily allow the truth to be stifled and the MPs who were constantly probing the government in the House had now clearly got wind of the attempt: 'If anybody picks up a succession of Hansards [written Parliamentary reports],' he threatened, 'they can gradually build the picture. And so it'll come out anyway.'

It came out rather sooner than the chief expected. When I first heard the taped conversation I had intended to use it only as unattributable background material to confirm our first story. No one would then be likely to know, I reflected, how a foreign government had attempted to deflect the blame from its own nationals in the face of all the facts, in a way that scorned the cause of air safety and did nothing to make a recurrence less likely. But possibly this was a rather sanctimonious response to the emotive arguments I had overheard and one that was best forgotten.

But I remained indignant. The telephone call had a more lasting impact: what impressed me was not only the chief's remorseless logic and his masterly exposition of the facts, but also his redoubtable defence of British airmanship in the face of an apparently calculated piece of political chicanery – and the more creditable because his advocacy was never intended to be heard. Faceless bureaucrats take a lot of stick, and in the aviation world the rank and file hardly regard themselves as allies. Here was an unsung heroic exception to disprove the rule, and a story too telling to miss.

Besides, more evidence of air traffic control inadequacies at Tenerife had since come to light from a Britannia Airways Boeing 737 that had been following the Dan-Air plane into the airport. It had been cleared to descend to the same level allotted to a third plane, a British Airtours 707 just ahead. The Britannia pilot said that a collision was averted only because, as he descended, he had happened to see the BA 707 flying 1,000 feet below him through a break in the clouds. Clearly the handling of Captain Whelan's flight was not an isolated event. Growing holiday traffic to the island was continuing to fly into danger while the politicians sat on the facts.

The next hurdle was to persuade Harold Evans, then editor of the *Sunday Times*, to use the sensitive eavesdrop material. I need not have worried. All he asked was: 'Are you sure it's genuine?' I was, and we published in full.

Official consternation followed. Characteristically, Harry Evans stonewalled the protests and he was truthfully able to refuse all inquiries about the source of the tape since neither of us knew it.

There was, however, one strange sequel. After our story appeared, a BBC producer pleaded for a copy of the tape on the basis that it was now spent ammunition to the *Sunday Times*. There was a working understanding about such mutual exchanges between us, and on that basis I agreed. The broadcast was for that evening and time was too short for me to make a copy of the tape cassette, although I had made a written transcript, so the BBC promised to make a copy and return the original to me forthwith. Shortly after, a broad-shouldered but laconic BBC messenger arrived to collect the tape and slid quickly out of the building.

I never got the tape back. Weeks passed, and when I inquired about the tape the BBC said there was no trace of it. So the identity of the burly messenger adds a jot more mystery to the equally unknown donor of the original tape that started the story. Was the giver also the taker? Any outward suspicions that stuck to me about an illicit source were more than offset by the glow of having rather brutally scooped the minister's eventual statement to the House of Commons on the Tenerife tragedy – as well as inspiring a number of questions to him that might otherwise never have been asked. And, thanks to the AAIB's chief, it is reasonable to hope that safety heights will be better observed and that the custom of conjuring up unapproved holding patterns for pilots in mid-air may not be lightly attempted in future.

Los Rodeos or Tenerife North airport has an unhappy history. It was, of course, the site of the world's worst aircraft accident in March 1977 when two Boeing 747 jumbos – a Dutch KLM and a Pan-Am –collided on the runway in fog killing 574 people. Here again misunderstandings between pilots and controllers played a major part in the crash, although the exact apportionment of blame is still a source of controversy. It is some comfort for future travellers to know that not long after the Dan-Air crash in 1980, the airport was closed to most airline traffic, which is now served by the modern Reina Sofia terminal that lies on the southern (and flatter) coastal tip of the island.

A safety league of airports

Each major crash at a terminal that has some known deficiencies usually brings a call for a blacklist of airports, and it is an exercise that can act as a spur to improvements in those regions susceptible to public criticism. But its value is limited because it at once leads into deeper waters. All airports have some limitations and it is the job of the pilots and the airlines to observe them. Operators are provided with a massive amount of mandatory data about every international airport with which they must conform, according to the terms of an airline's licence: besides local conditions and weather, the limitations on each flight vary with the type of plane and its equipment, the qualifications of its commander and other factors. To keep within the law, each flight has to be dove-tailed into these interacting constraints.

It follows that nobody *should* be flying into dangerous airports, in the sense that all airports are dangerous if the limitations – of all sorts – are not observed. Funchal airport in Madeira, for instance, may be perfectly safe for a 60-seat F-28 Fokker Fellowship jet, just adequate for landing a full-size B767 in good weather and into a nice headwind, but illegal in the hands of a junior pilot trying to land it with a cross-wind from a thunderstorm. So the burden of safety is pushed back from the operators at an airport – however grotty – onto the visiting airline. At least up to a point. The responsibility can be reversed, of course, when the airport generators have a habit of failing at night just as a plane may be committed to a landing, to plunge the runway into darkness and cut radio and landing aids at the crucial moment.

This is more than supposition. In 1978 183 people on board a DC-8 jet died when it crashed on its final approach to Sri Lanka's Colombo airport, after a succession of power failures that cut the navigational landing aid transmissions at the vital moment. While pilot associations immediately sought to blacklist the airport, deeper investigations revealed an astonishing record of near-accidents caused by sporadic failures of the airport's equipment. It came to light that less than a week before the crash, 600 passengers had circled above the airport in two jumbo jets and a DC-8 because the instrument landing system and two other radio aids had suddenly packed up. Days before this the airlines using Colombo had complained to Sri Lanka authorities about the dangers of the unreliable aids, but the row had only leaked into public knowledge through an assiduous local reporter with a line to London. The affair

106

led eventually to new equipment being installed at Colombo, but the scenario is far from unique.

But in the main such unforeseen breakdowns at airports are the exception, and it is the airlines and their crews who remain responsible for avoiding the known shortcomings of 'dangerous' airports. If there must be blacklists, a safety league of the airlines would make more sense than complaining about facilities at foreign fields. There have been two attempts known to me to publish airline safety comparisons and both have raised cries of defamation that led to profound legal turmoils. Any attempt to do this fairly is fraught with doubts, and at best it is not terribly meaningful, for one reason because the criteria of safety are bound to be partly subjective and so open to endless challenge, for another because airlines can change their places in the safety stakes during the time it takes to assemble and publish whatever statistics are thought to be relevant. While the industry swarms with safety statistics, one of the most lucid analyses, which provides a basis for some general conclusions, comes from the 1989 Lufthansa World Accident Survey reproduced in Appendix II.

Both these flaws reduce the value of the exercise, although it remains true that at any given time it is possible to separate the sheep from the goats. A friendly inquiry about the record of this or that airline can be most safely answered out of the side of the mouth in a reasonably informed manner (try an airline pilot, for instance), but it cannot be more than a hunch that is as reliable as the person who offers it. Woe betide anyone who hangs statistics around a guess or who publishes it as fact – too many innocents have been scorched by that flame. But the eternal question that remains is how are air travellers to choose the safer airline?

8 | Official secrecy

The Tenerife tapes came and went. Officialdom is, of course, mortified by such intrusions into the private processes of an accident investigation. Above and beyond the delicacy of inter-state negotiations, the legal eagles will be circling in wait for any fragments of information that emerge – whether by leak or otherwise – which may point to liability on the part of those involved. We have seen here that the airline Dan-Air was quick off the mark with its suit against the French authorities, and most major accidents launch a train of diverse legal actions, usually led by a parcel of claims for compensation by the families and dependants of the victims against the airline or the crew, the manufacturer of the plane or anyone else alleged to be at fault.

There may be sound policy reasons for bureaucrats to insist – as they do in Britain – that it is no part of their job to apportion blame, but undoubtedly one of the strongest motives for their circumspection is the fear of committing the unmentionable crime of exposing ministry officials to actions for defamation or compensation in the courts: government lawyers blanch at the rustle of a writ. The defence of privilege may not protect them, they fear, against a false accusation or implication of blame. Even a whisper of legal action would (and how often have they declared it) be terribly bad for their minister which, being translated from the mandarin, means terribly bad for their jobs. They are human and we should not blame them, but rather blame the timorous traditions of the system.

The little drama of the Tenerife tapes must have been mildly tiresome to them in this respect. One surmises that the Foreign Office perhaps uttered a tut-tut at the publication of such harsh words between the two countries at a time when there were renewed rumblings from Spain about the future of Gibraltar. But what no doubt seized the official mind was not so much how the leak had been obtained, but what *else* had been and might in future be overheard?

Although the dust has long since settled on the memory of this droll episode, my history of it may persuade them that it is not I who can answer that question. And they should hold hard before running a sword through the arras to skewer the unknown eavesdropper, in case he or she is on their side. It may be some small comfort to them to know, as I can vouch, that the general response to the full text of the telephone conversation as we published it was one of relief to know that there are red-blooded civil servants within the silent service who are willing to champion the cause of good airmanship and the future safety of international flights.

Inevitably its publication brought reproaches about journalistic irresponsibility, and they are valid to the extent that, as in all sensitive negotiations, the truth is more likely to emerge in confidential rather than open discussion. The trouble is that – as the recent spate of whistle-blowing memoirs have shown on a much broader canvas – once the veil of secrecy is drawn across one level of endeavour, the instincts of caution soon stretch it to the utmost to cover more than they should. Other motives for closing the doors on the public eye can then easily come into play. They may betray themselves in relatively trivial ways.

Differences between Britain and America

The AAIB of the Department of Transport in Britain publishes regular safety bulletins every few weeks which summarize recent air accidents or incidents, but it blanks out both the name of the airline involved and the names of the aircrew. This is a tiresome trick which can easily be circumvented by anyone with access to library references or with personal contacts in the industry. But the public is not to know, apparently in order to protect the image of the airline or operator. Serious accidents – a small proportion of those outlined in the bulletins – become the subject of detailed inquiries which are published with the identity of the airline and pilots in full. The inconsistency seems to reveal an itch not to tell until you must.

The United States parallel shows a more liberal attitude to disclosure. Airlines are named in interim bulletins, and the NTSB accident reports have no qualms about doling out forthright blame upon people, airlines or corporations where, in its opinion, this is deserved. For example, its findings on the Vieques, Puerto Rico, crash slams the FAA itself for 'ineffective surveillance' of the airline and the latter for 'failing to train the pilot adequately'. It identifies the pilot's 'failure to execute the emergency properly . . . and remove excess water from the tanks (before flight)', and it also spells out the airline's deliberate course of conduct in fudging many flight schedules to conceal its use of unqualified (and so cheaper) pilots. If such findings present a windfall to plaintiff lawyers, this is not allowed to inhibit the NTSB's verdict.

To be fair to the British AAIB, its tone has become noticeably sterner in recent years, although there is still a tendency for it to pull its punches. For example, a Nigerian DC-8 struck the tail of another aircraft parked well clear on the runway at London's Stansted airport as it made a low overshoot in bad visibility – so bad that the attempt to land was in itself illegal. It is charitable to say that the overshoot was also badly flown, as the plane slewed far to the left of the runway centre line. Although the DC-8 was substantially damaged, its crew managed to make a safe landing at Manchester. In all, it was not the kind of risky attempt expected of professionals with 58 passengers on board. But the AAIB didn't say that, as the NTSB might well have done.

The AAIB's 1984 findings of the Stansted accident that happened two years earlier does state that the crew failed to start the overshoot early enough – that is, at no lower than the prescribed decision height (1,000 feet in this case) if the runway cannot be seen. In fact the fog was down to as low as 100 feet above the airport. It adds that their flying technique was faulty. The report also states, as a fact, that the crew were told several times over the radio about the bad weather at Stansted. The remainder of its conclusions are courteous. Both pilots, it says, were unaware of the UK rules about banned approaches in bad weather and the data about this, which should have been carried in their manuals, was missing. This shortcoming was directed at the operator.

To anyone engaged in the game of flying, the implications of blame are clear enough, particularly so if they are familiar with the AAIB's gentlemanly approach. But to members of the public – and the report is after all a public document – it reads as not much more than a mild enough remonstrance for some technical shortcomings. Yet to airmen, the crew's errors are painful – and of a kind that might be compared to a charge of dangerous driving on the roads, and could be expected to earn magisterial strictures severe enough to be reported in the press. The AAIB does not say whether the pilot of the flight from Lagos had landed at Stansted before. If he had, he should have known the rules; if he had not, his attempt to land at a strange airport in thick fog in defiance of all the canons of good airmanship calls for the clearest public warning.

If the future safety of air passengers involves the exposure of blame – if only *pour encourager les autres* – surely it should not be shirked. The passengers themselves, both as paymasters and as potential victims, must have a preferential right to know and *understand* the realities in open language. The fact that technicalities may be beyond them makes it all the more necessary that the verdict should include a frank expression of expert opinion which may now only be inferred by reading between the lines. If this should throw blame on individuals or groups, then their *amour propre* takes second place to the whole truth.

Avoiding the attribution of blame is a professional article of faith sincerely held by the AAIB inspectorate, and at first flush it seems a generous one. But it breaks down in the face of experience. If human fault is a cause, there can be no less necessity to name it than there can be for any other cause. Since we are talking about human conduct, 'blame' must naturally follow because it infers some wrongful and avoidable behaviour. It would be poetic rather than sane to blame a seagull for entering a jet engine during the take-off run, but useful to blame the pilot if he failed to shut down the engine and cope with the ensuing emergency. When life and limb are at stake, naming names unhappily becomes a necessity if by doing so it is likely to reduce the chances of a recurrence. Remember Captain Dorman, the pilot of the infamous flight to Basle: had the earlier shortcomings in his career been on record, he would not have been at the controls of the Vanguard on that day. The engineer who botched the electrical repairs to the vital navigational instruments of his plane remains faceless and lived on for a while to continue his trade. And the French report of the Paris DC-10 crash miserably failed to pinpoint the human chain of errors that continued to threaten the lives of passengers until the focus of the infection was located deeper within the industry.

The canon that accident reports should not blame people is not only widely accepted, but it is also a part of the letter of the law

111

imposed upon officials. The UK accident regulations lay down that the purpose of investigations is to determine the causes 'with a view to the preservation of life and the avoidance of accidents in the future; it is not the purpose to apportion blame or liability'. Any strictures on this philosophy cannot therefore be laid at the door of the AAIB inspectors, however much they welcome an exclusion that keeps them out of the more contentious world in which their US counterparts must operate.

Such criticism as there is of the AAIB is confined to the delays in publishing its accident reports. For example, the relatives of victims and others concerned in the Manchester airport fire disaster of August 1985 had to wait nearly four years before the final verdict appeared. Here again, the inspectorate can point to the rules that require them to submit drafts of reports to all those who are affected by it and await their comments before the final version may be published.

Within weeks of the eventual release in March 1989 of the Manchester accident report in Britain came a US report of a major accident that had happened only just over a year before, on 28 April 1988. The cause – the most massive structural failure ever survived by an airliner – astonished the world and posed questions that were more far-reaching than those raised by the Manchester affair. A large section of the roof of an Aloha Airlines Boeing 737 tore away while it was cruising at 24,000 feet on a flight to Honolulu, yet it managed to make an emergency landing at a nearby Hawaiian island. A stewardess was sucked out of the gaping hole to her death during flight and 69 of the 95 occupants on board were injured. The Boeing, nineteen years old, was on its 89,681st flight.

The May 1989 report of the accident by the US National Transportation Safety Board blamed the airline for failing to maintain the aircraft properly and blamed the Federal Aviation Administration for failing to uncover the airline's defective maintenance programme. Boeing escaped direct blame: but the Board noted that although the manufacturer had warned the airline that its four oldest B 737s needed repair, it had stopped short of condemning them as unsafe.

The British AAIB report of the Manchester tragedy (p. 268) puts the main cause down to the failure of the left engine, followed by a detailed analysis of the factual circumstances and a list of .

recommendations. There is no overt criticism of the human agencies involved in the story, nor is it easy to infer from the text whether or not anyone was to blame. The inspectorate has, as always, kept well within the letter of the law.

But how far is this policy of caution an honest attempt at gentlemanly conduct, or should it be seen more as an exercise in damage limitation inspired by the government's in-house lawyers? Certainly much of the heat of criticism is dispersed by the consultative process. But as a Member of Parliament declared during the long wait for the Manchester accident report to appear, the internal reason for the delay was the fear of legal claims for compensation that might be founded on even the implications of a factual report: manufacturers, airline and personnel might all become vulnerable. Alas, they might: as the Americans say, if that's how the cookie crumbles, so be it. If safety is to have first priority, other troublesome consequences must take second place.

Here is the nub of the shortcomings of accident reporting in the UK, with all the constraints and delays that surround investigations. The less inhibited American ethic which expects blame to be seen to fall where it should – whether on man or machine – produces more rigorous sanctions against lapses in safety standards. Human error remains by far the commonest cause of accidents, so that to delete this as a matter for legitimate comment must reduce the value of inquiry to that extent. You may point to a faulty instrument or a wing that drops off with relative impunity, but it takes more courage to point to a man – be he pilot, airline manager or engineer – who has good lawyers behind him. Remember Walter Bagehot's advice to his seven-year-old nephew who was trying to break into his boiled egg at table: 'Strike it hard, Guy – it has no friends.'

The British rationale for not naming names appears to be that where it is necessary, disciplinary action is indeed pursued behind the scenes and that this is an equally effective and more civilized manner of conducting affairs. But how do we know, and should we not know? Private admonitions or even firings may deal with the case in point, but they are no deterrent to others not in the know. Aviation is a harsh environment where the sanction of publicity may be a regrettable necessity. No one will deny that the dangers of secrecy are heavily loaded in matters of transport safety, and it is here that the call for more open government sounds most clearly. But the roots of that affliction have been shown to grow deep indeed, from the classics of Watergate in the USA to the more recent spillings of Peter Wright's *Spycatcher* in Britain. It is now less easy to be rid of the thought that a common motive underlies these as well as the more prosaic examples of official reticence. The temptation to resort to executive secrecy remains the same – the desire to disarm criticism and effective public scrutiny. A couple of wisps from recent history provide a pertinent comparison between the US scene and the British system.

The first incident happened above New York's Kennedy airport when a British Airways B 707 nearly collided with a light plane in the summer of 1980. In Britain, it revived another of the periodic outbursts of alarm about the mixing of small planes and airliners near busy American airports. FAA spokesmen, responding to inquiries from London only hours after the event, were quite uninhibited and confirmed that an airmiss report had been filed by the Boeing pilot. This they quoted in full over the telephone, adding the unsolicited offer of further data as and when it came in.

Americans might see this as a normal and unexceptional response: all of those with a role in preventing a recurrence get the facts as a proper and immediate contribution to air safety. (In parenthesis, it ought to be said that this was no exception: covering American stories is jam to British journalists – not only can they enjoy a frank exchange on first-name terms with most corporation chiefs at the drop of a hat, but most times the boss is found to be at his desk during working hours. In contrast, Britons are familiar with those official spokespersons who defend themselves behind the twin shields of absence and a fear of going on record; the functionaries who fade out on a promise to ring you back, but seldom do so.)

A week after the New York airmiss was dutifully reported in detail, an excited Scottish witness telephoned to report that he had seen a wandering German bomber nearly collide with an airliner as it came in to land at Aberdeen's busy oilfield air terminal in Scotland. He told how 25 frightened passengers had disembarked from the civil plane demanding to know what it was that had nearly hit them. Officials were not going to enlighten them.

The CAA told inquirers, correctly, that it was unlawful for their officials even to confirm or deny not only the 'incident' itself, but whether or not an airmiss report had been filed by either pilot. It had:

114

but the facts that led to the near-disaster had to emerge from other sources. The presence of a military plane had had nothing to do with the clampdown by the authorities.

Seasoned inquirers know that reports of airmisses between civil planes are guarded as though they *were* a matter of national security. The conventional reason given for this is an administrative one: all airmisses are dealt with by a joint civil and military working committee, with the result that the otherwise understandable security silence about military mishaps must be stretched to cover civil airmiss reports as well. Presumably, when military data is present in the same committee room with civilians, all are gagged by the Draconian powers of the Official Secrets Act. The tacit reasons for classifying such information can be deduced from what has already been said: it makes for an easier life and keeps the lawyers guessing. And nobody wants to upset the airline trade by endorsing newspaper reports that only alarm the public unnecessarily.

It was only when public anxiety about the recent series of airmisses both at home and abroad reached Parliament, so that the rules about the suppression of airmiss reports were dragged into the light of day, that the politicians were forced at last to declassify them (in 1988).

As a rider to these two airmiss reports, the different manner in which they were handled in the USA and the UK reveals more than contrasting official attitudes. Questions about the Aberdeen incident were met by a familiar wall of silence even from unofficial desks and other observers who had no visible motive for such discretion, other than an inherent distaste for becoming involved even in a matter of obvious public concern. Luckily there are a few exceptional spirits who are disposed to whistle-blow and so prove the rule. By contrast, the American impulse is – far from keeping the news to themselves – to reach for the telephone.

Watchdogs

Nobody supposes that air travellers can ever be given a day-to-day reckoning about the safety of this or that airline or flight, or all the insider technical knowledge that bears on flight safety. Most would not want it or wish to understand it. What they are entitled to expect is a watchful public eye to monitor the scene on their behalf – the prime job of government regulatory authorities such as the FAA and

CAA under their respective departments of transportation or transport.

However well they succeed in this role, they fail us when they do their work behind closed doors and so, insidiously, slip into that too-cosy nexus with those whom they are supposed to be policing. In that sort of climate it is no surprise that their public utterances are limited to carefully sanitized press statements which tell the public no more than it is likely or able to find out by other means. Bureaucrats know that the public has a short memory and that it takes time to unravel the whole truth. It is their hope that once the initial shock of an accident has faded, there will be little news value left in the detailed post-mortems that may follow in the pages of official accident reports or *post hoc* investigations by the media.

Happily that hope is occasionally defeated and the public interest aroused when the whole truth emerges. Much more importantly, this in itself can become the strongest sanction against official secrecy by teaching the salutary lesson that it is safer to tell all rather than risk the rap of being found out later.

As their main defence, customers have to rely almost exclusively on paternalistic bureaucrats. However dutiful they may be – and in Britain we are blessed with those that rank among the highest worldwide (few would quarrel with a listing that puts the UK, Holland, West Germany and the USA in the top bracket) – enough has already been said to show that bureaucracy as a whole is both slow and fallible. Government officials, dedicated as they are, are subject to their political masters and to their departmental lawyers.

In a search for heroes in the world of aviation, my prizes would go unhesitatingly to the British Air Accidents Investigation Branch (now labelled for administrative convenience under the Department of Transport), and to the US National Transportation Safety Board in Washington. As safety watchdogs, both have achieved and kept enough real independence from their government paymasters to allow them to fight their corner against the powerful law-making and executive agencies – the Civil Aviation Authority and the Federal Aviation Administration respectively – to whom the agencies report.

But however loudly these watchdogs may bark, it has been shown how the higher bureaucracy above them too often turns a deaf ear. The job of both the CAA and the FAA is, of course, to balance the commercial needs of the industry with the protection of the airline customer. As parts of the state machine, they must also work to other imperatives such as the balance of payments returns from tourism, care for the jobs and health of the manufacturing sector, as well as the diplomatic tit-for-tat bargaining over reciprocal entry rights between national and foreign airlines.

All these diverse duties and pressures, attended to by their large armies of government servants, ensure that safety is but one concern of many: and it is clear from their various roles that safety – and the cost of maintaining and extending it – collides head-on with these other key functions. It is a flaw that at last may be remedied following government hints in June 1989 that the policing and law-making functions of the CAA may be separated. The necessary business of hammering out these conflicting duties into a policy of compromise is, of course, done behind closed doors. The public will not know if or how far the frontiers of safety may be dented in the process, apart from guesswork that may be sometimes inspired by a leaky official. Nobody will deny that there are matters that must be discussed in confidence. The trouble is that those who wish to keep them off the record are the same as those who decide what shall and what shall not be kept from the public. As we shall see, the instinct towards expanding these secret empires is not confined to government agencies. The habit of silence on airline safety matters brings some vivid and tragic consequences in its train.

9 | The secrets of Nairobi

In the five years after the first Boeing 747 jumbo went into service with Pan American in January 1970, it had already established itself as probably the safest and most rugged flying machine in airline history. Nor was its safety record blemished until the first fatal B 747 crash came at Nairobi in November 1974, a story of official secrecy at its most dangerous.

The crash

Thankfully there were only 157 people on board Lufthansa flight 540/19 as its co-pilot taxied it out for take-off in the early morning sunlight. There had been the usual exchange of some passengers during a brief refuelling stop on its long southerly trip from Frankfurt, Germany, to Johannesburg in South Africa. Hans-Joachim Schacke, the 35-year-old ex-German air force pilot who was now at the controls, was familiar enough with Nairobi airport from his twelve previous flights there. Beside him sat Captain Christian Krack, who had also started his career in the air force and, at 53, now had three times the flying experience of the younger pilot who was due to fly the second leg of the journey out of Nairobi. With the pre-take-off checks completed, Schacke lined the jumbo up on the runway and eased the four engines' thrust levers forward. According to routine, the captain on his left and the flight engineer sitting behind him were both monitoring the instruments as the plane gathered speed without demur on what appeared to be a perfectly normal take-off.

Seconds after the plane lifted off at the prescribed speed of 145 knots (165 m.p.h.), the crew felt vibrations and buffeting. Krack instantly confirmed with the engineer that the engines were delivering normal climbing power. Next, he thought the shaking could be due to some imbalances in the huge main wheels which were still spinning, and he retracted the undercarriage. But meanwhile the plane's acceleration ceased abruptly and it failed to reach the normal target airspeed needed to achieve the initial climb-out. Schacke instinctively flattened his angle of climb to allow the speed to build up, but still the jumbo failed to respond and sagged into a gradual descent as it passed only 100 feet above the airport perimeter with the airspeed now decaying to 140 knots (161 m.p.h.). At that instant, the stick-shaker operated – an insistent automatic device that warns pilots when the airspeed falls below safety limits. Krack, as commander of the flight, seized the control column to lower the nose further to gain speed, but he saw they were now too near the ground to do so. Meanwhile Schacke realized they were going to hit the ground and closed all four engine throttles. In the few fraught seconds since lift-off, neither of the two men had time enough to eliminate the possible reasons for the jumbo's refusal to fly and so to isolate and remedy the cause of the impending disaster.

Ironically, there was another pilot on board in the passenger cabin who knew what was wrong, but he was powerless to prevent the coming crash. He was seated in the left-hand window seat near the wings in row 16. As a former airline pilot, when he felt the plane vibrate after lift-off he looked out at the wings. To his horror he saw that the leading-edge flaps had not been extended. This must be done before take-off, and it was the lack of the essential extra lift that they should have provided that was the simple cause of the vain struggle for survival that was going on unseen behind the closed cabin doors to the flight deck.

The first impact came when the tail of the B 747 struck the ground 1,120 metres from the end of the runway, causing the plane to bounce a few feet into the air before it sank again to smash into a raised roadside bank. The tail disintengrated but the main section skidded on for a quarter of a mile, turning as it did so, finally coming to rest facing in the opposite direction. Fire broke out on the left wing, followed by an explosion that spread flames to the fuselage. The cockpit doors had jammed but the crew escaped through holes in the fuselage as fire engulfed the cabin. Fifty-five passengers and four cabin crew died in the inferno; another 54 were injured before they could be rescued. Hours later the sight of the still-smouldering wreck of the first Boeing jumbo to crash showed to the world that not even the proudest plane is proof against the perils of the air.

Why had its flight crew failed to extend those leading-edge flaps? The answer is a grisly one. The accident need never have happened. That week the *Sunday Times* carried an article I wrote under the headline, 'Jumbo crash: the safety device Lufthansa does not know about'. Its first paragraph gives the gist of the answer to the question and opens a sad story of regret and recrimination:

> The Jumbo jet crash in Nairobi last Wednesday was caused by the failure of the flaps on the forward edge of the wings. British Airways suffered a similar failure two years ago, but the safety modification they insisted on then was not passed on by Boeing to Lufthansa, the operators of the Nairobi jet.

The safety systems of the B 747

The events so far revealed lifted the curtain on an astonishing story of untold warnings and seemingly persistent neglect that, to be properly understood, calls for some insight into the safety systems built into the B 747. Most passengers will have seen the trailing-edge (TE) flaps that seem to extude from the rear of the wings in a downward curve a few minutes before their plane comes in to land: the same flaps are extended before take-off and may be seen withdrawing into the rear wing as the plane climbs away from the airport. They provide the extra lift that is needed to fly at slow take-off and landing speeds.

The amount of lift generated is proportional to speed as well as the size of the wing area. As jets are designed to cruise at speeds of over 500 m.p.h., the slim wings give enough lift at high speed but not enough at airspeeds around 150 m.p.h. – the kind of speeds necessary in the transition from air to ground and vice versa. Flaps, which are traditionally extended from the rear of the wings, provide the answer and permit the widest range of safe flying speeds. But as planes get bigger, heavier and faster, their take-off speeds would need to increase proportionately to gain the added lift they need. And as tyres and wheels are already touching safety limits at speeds of 150 m.p.h.-plus as they race over the tarmac, designers are constantly seeking more efficient flaps to extract more lift rather than to obtain it from higher and higher land speeds.

Some people also may have noticed that the *front* edge of the wings

1 Survivors are helped away from the DC10 fire in which
55 died at Malaga airport on 13 September 1982

2 The burnt-out fuselage at Malaga

3, 4, 5 *below and opposite*
French police search for the remains of the 346
who died in the Turkish DC10 that plunged into the
forest at 497 m.p.h. near Paris on 3 March 1974

6 The final destination of the Axbridge Ladies' Guild
in the Swiss Mountains where 108 died
on 10 April 1973

7 The charred remains from the Swissair DC-8
that overshot the runway at Hellinikon airport, Athens,
on 7 October 1979

8 Manchester airport 22 August 1985: 55 died in the fire
after the BA B737 twinjet aborted take-off

9a, b Sharon Ford and Jacqueline Urbanski,
the two British Airways stewardesses who gave
their lives in helping others to escape

10, 11 *above and opposite*
The British Midland B737 twinjet that crashed
on to the M1 motorway on 8 January 1989, killing 47 people

12 The tail of the Air Florida B737 twinjet
is hoisted from the icy Potomac river
after failing to climb out from Washington
airport on 18 January 1989:
5 out of the 79 aboard survived

appear to separate and form a sort of double curved edge leaving a gap between. These leading-edge (LE) flaps – which are now used by most large aircraft and are arranged in 26 sections along the wings of a B 747 – are a cunning aerodynamic device that generates a lot more lift with the least 'drag'. The lift provided by both TE and LE flaps allows the lowest possible flying speeds for take-off and landing and, when all are tucked into the wings, the highest cruising speed.

Pilots have the choice of four or five settings to vary the total amount of flap to be extended as may be required in various phases of flight and to suit landing and take-off conditions (aircraft weight, length of runway, weather conditions and engine power all govern the amount of lift needed on a particular day). Here we touch one of the sequential factors that contributed to the Nairobi crash: *a single lever* controls and harmonizes groupings of both sets of flaps. An equally vital point that concerns us is that while the TE groups are powered hydraulically, the LE flaps are driven pneumatically from air 'bled' from the engines.

Lufthansa required crews to close the air valves to the pneumatic system while the engines were being started, and their pre-take-off checklist reminded them to turn those valves on again before the flaps were selected. The first simple point is that the Nairobi crew for some reason seem to have omitted this vital item. But Boeing (like other designers) fitted two built-in safeguards against such simple oversights by crews. In front of the pilots a warning light shows green when the LE flaps are properly extended and on the flight engineer's panel there is a more comprehensive display of warning lights giving a separate green for each of eight groups of the 26 sections on the wings. Both for pilot and engineer, it follows that no greens means no flaps extended.

According to the investigators, all of the crew seemed to have failed to check these greens: they add that even double-filament bulbs and electrical systems can go wrong, although this they thought to have been unlikely. There is, however, another point that helps to explain the crew's oversight: the green light that warns the pilots of the state of the LE flaps is placed close to three other green lights that illuminate when the undercarriage is down and secure (as of course it was at take-off). A failure – sooner or later – to spot the difference between a group of three and of four greens is more understandable. Another confusing factor was that Lufthansa and some other European carriers had just changed their engine start-up drills to allow this to be done with the LE pneumatic valves closed, contrary to previous practice.

But still the Boeing designers were watchful. For pilots of smaller planes, the moment of truth arrives just before the throttles are opened for the take-off run. However carefully safety checklists may have been done, there remains a final shortlist of vital precautions burnt into the memory that, if forgotten, could spell instant disaster after committal to a take-off. These last-minute prompts ask: Flaps set? Engine indications OK? Brakes off? Runway heading correct (compass direction)? (The last shows that at night-time it is easy enough to line up on the wrong runway at big airports.)

Airliners combine such vital 'no-go' items – which are rather different on a big jet – into a single automatic 'take-off configuration warning' system (TOCW). Simply put, a klaxon blares out on the flight deck if any vital item has not been correctly set. The correct setting of the flaps is of course one of these, as the Nairobi crash demonstrates.

Because both sets of flaps are controlled by one lever, the klaxon circuit was wired to alert the crew only about the position of the TE flaps, on the logic that the LE flaps must move out in harmony when the TE flaps are set. But if the pneumatic power for the LE flaps was cut off for some reason, the system would assume both sets had been extended when in fact the LE had not. In other words, the failure of the forward LE flaps to extend would not, on its own, trigger the klaxon.

The warning klaxon is wired to sound as the third engine (inboard right wing) thrust lever is advanced beyond 50 per cent of its travel, if any item is wrongly set or out of order. Had the LE flaps been treated as a separate vital item, therefore, the klaxon would have warned Schacke to abort the take-off as he pushed the throttles forward and before the plane began to gather speed. Had no one foreseen the Achilles' heel of the system – forgotten pneumatic power valves that were still firmly closed? They had not.

Neglected warning

The investigations that followed revealed that the leading-edge flaps had been overlooked on no less than eight previous take-offs by other

B 747 operators. The incidents were duly reported to the FAA but they were ignored. The cost of that inaction was to be a high one. It was evident that the warning system could be defeated by human errors made by even the most highly trained airmen. The dangers were compounded because mechanical safety actions that were once performed step-by-step by pilots breed a reliance on the seemingly foolproof machine. Mistakes that happen in other phases of flight may be corrected without serious results, but omissions in the 'vital actions', as they are often called, are more than likely to be lethal.

It says much for the design of the plane that none of these earlier mistakes proved to be fatal: Nairobi airport is 5,327 feet above sea level and the thinner air at that altitude calls for the maximum lift (compare the airport at Denver, Colorado, which as it happens is just three feet higher). Those conditions proved to be too much even for the safety margins built into the B 747 when it was asked to wing its way into the air without the aid of its powerful forward flaps. (The point is academic, but even then the crew in fact had unused reserves of engine power which might have saved them if they had been able to identify the cause of their troubles early enough.)

But as long as two years before Nairobi another scare caused by a similar oversight was treated very differently. It was dutifully reported by the crew of a BOAC (later British Airways) flight. In this case only half the LE flaps on each wing had been extended before take-off, this time because an electric circuit breaker had been left in the tripped (or off) position after an engineering maintenance check. The company's chief safety investigator, John Boulding, wanted to know why the crew had not been alerted by the TOCW warning system. His team discovered, much to everyone's surprise, that the LE flap system was simply not connected to the TOCW device. Further, they foresaw that the green warning light circuits were wired in such a way that, under different conditions, it would be possible for pilots to take off without any of the LE flaps extended – probably with fatal results.

'This,' says Boulding, 'was an intolerable situation.' The warning system was designed to alert crews to unsafe configuration before take-off, and yet here on the B 747 some of the most vital items, the LE flaps, were not connected to it. Boulding's team immediately asked the airline engineers to devise a remedy – a simple matter of wiring in the LE flap circuit. This received the approval of Boeing,

123

who issued a maintenance service bulletin. But, as will be seen shortly, this bulletin was only issued for BOAC and Aer Lingus B 747 fleets. (The Irish jumbos were at that time maintained by BOAC.)

While this paperwork was still in the pipeline, all BOAC crews were immediately warned about the potential dangers in the system and of the need to double-check their LE flaps. Meanwhile the saga was reported to the Civil Aviation Authority in London in the knowledge that it in turn would alert the Federal Aviation Administration in the United States, which had certified the model as airworthy in the homeland of most B 747 fleets. But Boulding and his airline were dismayed when they heard the FAA's response. It declared the BOAC modification to be unnecessary: the existing crew safety drills, it said, were adequate, nor was the TOCW warning system mandatory equipment under FAA rules. Neither Boulding nor his airline were disposed to accept that view: if a warning system was provided for crews to rely upon, it must be foolproof. Indeed, a flaw that omitted such a vital warning as the position of the LE flaps made it worse than useless. They argued that without the remedy that they had devised, it would be safer to scrap the TOCW altogether and fly without it. Crews would then revert to item-by-item checks, rather than build false confidence in an automatic device that was meant to do the job for them.

Some time later at an airline safety meeting in London, and five months before the Nairobi crash, a KLM safety officer revealed that a B 747 training flight had taken off with its LE flaps retracted, but its crew had spotted this in time to avert a disaster. When the airline discussed the incident with Boeing in Seattle, the manufacturers referred KLM to the bulletin they had issued after the BOAC scare. It was then that Boulding realized that the bulletin had only reached BOAC and Aer Lingus: KLM had known nothing of it. Boulding was now disturbed that nothing seemed to have been done to alert B 747 operators worldwide in the two years that had passed since the danger had first revealed itself. A second news article I wrote in the *Sunday Times* picks up the story from that point. It began:

The unexplained loss of three vital letters from British Airways, warning the Civil Aviation Authority about hazards in handling the B 747 Jumbo, is causing an urgent review of safety procedures. They were written 3 months before the first fatal accident to a B 747 and predicted the events which led to the tragedy [at Nairobi]. John Boulding, a BA air safety adviser, wrote the first letter to the CAA in August 1974 but it went astray. A copy was also lost, as was a third letter. A spokesman for the Department of Trade – which is responsible for airworthiness standards – confirmed on Wednesday that an urgent review had been ordered between the CAA and the airlines 'to ensure that there would be no future problems between the CAA and the airlines about the handling of safety communications'.

Boulding's letters had listed a whole series of carbon-copy errors that had stemmed from the same cause and described the action he had taken to prevent them at British Airways. He went on to add a warning that was to become sadly prophetic: 'The industry has already had two close shaves and the next Operator might not be so fortunate.' He urged the CAA to persuade the US authorities to make his modification a mandatory requirement for all B 747s worldwide.

Puzzled by the CAA's initial failure to reply, when Boulding later met the person to whom he had written the official denied that he had seen the letter. So Boulding shot off a copy to the same CAA fellow, but again nothing but silence ensued. When he finally called in person at the Authority's offices to check that his second letter had been received, the official again denied that it had reached his desk. So far, this kind of bureaucratic episode is perhaps too familiar to astonish. But besides being a seasoned tactician, Boulding had become thoroughly uneasy – it was now only weeks before the Nairobi crash – and he sent a third copy of his letter to the official's home address. This time, he assumed, the shaft must have hit the target.

When the Nairobi crash happened, Boulding contacted the CAA and demanded to know what had been done about his repeated warnings, firm in the belief that at least his last attempt had entered the system. It was then that he was told that the CAA had no record of any of them – not even from the home of the official, who denied having received it. When he demanded an explanation from its senior echelons, the story broke in the media and forced the CAA to announce its 'urgent review of safety procedures'. As the cudgels cracked in the row that ensued, the British Airline Pilots' Association publicly accused the CAA, the FAA and Boeing of failing to heed 'a fault of manifestly lethal potential' until an accident had happened – words that smack of genuine outrage from a body not greatly given to hollering about its professional worries in public.

In the event, the product of the CAA's urgent inquiry was to be no more than an admonitory reminder sent out to all British airlines, which told them that it was *their* duty to ensure that letters reached the Authority's officials. The insolence of office, it seems, has no bounds. A generous interpretation of the CAA's part in the affair may be that, if all three letters had in fact gone astray, its only safe exit from the altercation was a headmasterly wag of the finger. It may be hard to believe that, as it were, lightning can strike three times in the same place, but it becomes more credible seen in the context of the vast amount of paperwork circulating between airlines and the officials who must sift the grain from the chaff. If there had indeed been a deliberate cover-up, it may have followed the rule that to admit one would inspire suspicions of others. In other words, to adapt the old mandarin saw – let us not admit today, lest we be asked to admit more tomorrow.

As a rider to this revealing history of how a simple but lethal flaw comes to be hidden in such a highly sophisticated piece of machinery, the invention of leading-edge flaps (or slats as they are sometimes called) may interest the curious. The three disasters in the early 1950s to the first passenger jet, the British Comet 1, revealed that the wings tended to stall at the high angles of attack needed for take-off. It led British designers to exploit and improve on an idea from the past – leading-edge slats. Modern aerodynamics turned them into an extremely efficient method of producing high lift at low speeds. The British fitted them to their Tridents and VC 10s, while the first models of the competing Boeing 707s at first used only a small section, to be enlarged on later types. Their piecemeal introduction may explain why designers tended to regard them as 'add-on' devices. For example, while pilots are provided with gauges to indicate the position of the traditional trailing edge flaps, none is fitted for the LE flaps. At that time, the pilot's only safeguard was the single green warning light, coupled with the confidence – misplaced as we now know – that if the TE flap indicator shows that they are extended, the LE flaps must follow suit.

Needless to say, *after* the public outcry about the Nairobi disaster the FAA issued a mandatory modification to the TOCW system on all B 747 aircraft. Its design was rather more complex than the original fix, but the principle was similar. Boulding and his airline are the only ones who emerge with any credit from a story of secrecy, sins of omission and the sorrow of many needless deaths.

The cost of implementing safety improvements is a powerful cause of procrastination. As John Boulding commented, 'Once you've paid in dead bodies, money suddenly ceases to be a problem.'

10 | The role of the media

All this secrecy makes the role of the media even more crucial. Journalists who investigate air crashes are wise to the fact that lawyers and bureaucrats, however quick-witted, are not so fast on their feet. A crash scenario usually begins with a flood of information from the scene told by survivors, eye-witnesses, and pilots or airline colleagues of those involved in the accident. Experts in the industry will continue to doubt it, but is a matter of personal record that in nine cases out of ten, at this early stage a person with a knowledge of the industry can isolate the main causes that will eventually appear in the official accident report many months after the event. By that time, of course, every detail will have been meticulously scrutinized and proven before it is allowed to see the light of day. The official verdict can be held up for a year or more if the investigation raises complex issues or political overtones – foreign crashes, for example. (The State in which an accident happens has the right to conduct the inquiry, usually 'assisted' by officials from the State in which the aircraft is registered or manufactured.)

A score of nine out of ten may sound a high claim, but the odds given are in fact conservative. Success in piecing together the jigsaw of an accident comes not so much from the skill of the newsman but from the genuine eagerness of his informants to help solve the causes. Some, for instance, may be friendly pilots who confide that 'we've had that happen to us before', while others may want to spill the beans about a known defect or some unsafe practice. Most contentious stories present at least two conflicting factions that are each only too ready to tell what they know about the other. Other diverse sources range from expert contacts in the field, passengers and unsolicited eye-witnesses, lawyers acting for 'the other side' anxious to know what you know, and perhaps even politicians looking for a scandal and so willing to barter insider knowledge for the latest titbit as the story develops.

As facts funnel in from a wide spectrum of sources, it is not surprising that the journalist can come to realize that he has a wider perspective in the round than any of his informants, although without, of course, any particular merit on his part.

The charge that newspapers indulge in instant and precipitate judgments about the cause of aircrashes are understandable when they come from expert sources, and it would be absurd to suppose that there can be any competition between the authority of official accident reports and the first reactions of a reporter. Yet however technically profound their final verdict may be, the professional investigators face a number of fairly obvious constraints that are, however, sometimes overlooked. Primarily, they deal with mechanical failures and the human element in the quasi-medical sense – stress, medication, duty hours and the like.

However crucial they may be to the accident, anything touching on politics either at home or abroad is naturally beyond their civil service remit, and far more so, any critical part played by government departments or – perish the thought – ministers of the Crown. Journalists may be limited by time and knowledge, but their ambit is unrestricted. Even in the light of the few accidents already cited, it can be seen that some show primary causes that lie in a realm prohibited to government servants.

Although newspaper sources need to be evaluated, of course – there are many with an axe to grind – the amount of accurate information that pours in after a crash is remarkable. Whether or not jumping the gun in true newspaper fashion is to be deplored as irresponsible, or to be seen as fair game in the public interest, is something to consider later. Two or three days after the first news of a crash, the lawyers and government agencies will move in and gag the witnesses. Fortunately they usually forget a few.

The 1974 Paris DC-10 crash

The legal battle for secrecy reached untold heights after the 1974 Paris DC-10 crash (see pp. 53–6). There were no survivors, and the dependants of those killed had taken their claims to a Los Angeles court with the hope that the strength of their case would lead to an early out-of-court settlement. The defendants included McDonnell Douglas, the manufacturers, and the Turkish airline THY.

Presiding over the court was Judge Pierson Hall, who had agreed to an appeal by the defendants that the evidence and all the internal documents that were being put before the court should remain confidential until the case was finished. The claimants' lawyer had, of course, free access to the documents, which would otherwise become public knowledge as they were introduced into court. Two *Sunday Times* journalists were working closely with the claimants' lawyers under an agreement to exchange information and provide mutual help. The judge's ruling in favour of the defendants meant that the journalists were debarred from studying the bulk of the case-papers.

A bizarre episode of skulduggery ensued. One of the *Sunday Times* team, Paul Eddy, managed to pack the entire 50,000 pages of the wanted documents into a suitcase late at night and carry it out to a deserted office car park. But the catch of his car boot had jammed and his attempts to slam it shut roused a security guard. Luckily he turned out to be a friendly character who became more interested in helping Eddy to tie down the boot with a piece of string than in its contents. The papers were duly photocopied and the originals returned overnight.

There was no intention to publish any part of them at the time. The aim of the two reporters was to build a detailed examination of the faulty design of the plane's cargo door – known to be the prime cause of the crash – in their role partly as unofficial researchers for the claimants' lawyers and partly as newspaper investigators working to provide background to a detailed reconstruction of events that was to follow. As one of the lawyers, Stuart Speiser, claimed later, this alliance led to one of the most minute legal examinations of an aircraft's history ever made.*

It started for me on the day of the crash, Sunday 3 March 1974, while I was flying a light plane not far from the cliffs of Dover in Kent. My two radios were tuned one to base and the other to the regional London air traffic control centre. The usual staccato of radio calls between other planes and the ground – always entirely depersonalized, brisk and economical – was interrupted by a familiar voice speaking from my base to a colleague in the air, but in a most unfamiliar style.

* *Good Times, Bad Times*, Harold Evans, Weidenfeld & Nicolson, 1983.

'There's been a terrible crash near Paris – I've just come in and the air's full of it – no, no details yet . . .'

It was the voice of my ex-flying instructor, Don Nay, and I knew that only the shock of the news could let him abuse the airwaves with non-professional chatter. I changed to a Paris radio frequency and found that Don was right. Through an excited stream of French I first caught the word 'Creil' – a military aerodrome to the north of what is now Charles de Gaulle airport (at that time not yet open). For the first time I heard the name 'Ermenonville' mentioned. A glance at my map showed me the forest of that name and a small airfield called Le Plessis-Belleville lying on its edge.

My first impulse was to head straight across the Channel to become the first foreign journalist at the site. I tussled with a divided loyalty as a newsman – I had only recently turned from lawyer to the trade – and an aviator. Being on a domestic flight, I had no passport, prior customs clearance or flight planning permission to enter French airspace, and the consequences of an illicit swoop into Le Plessis-Belleville (not a customs airport) were daunting. I pictured gendarmes brought from the village impounding the aircraft, if not myself, and a rare old Gallic row that could well impede any journalistic enterprises. Nor did I want to throw away my tender but unblemished pilot's record. I persuaded myself that I would be more use in the London office and flew back to base, feeling the better as a pilot and the worse as a reporter.

Hearing Don Nay's voice over the radio struck an ironic note. Since the days of flight training he had become an airline pilot with Invicta Airlines. It was one of its Vanguard turbo-prop planes that had flown the Axminster Ladies' Guild to their fate in Switzerland a year before, and soon after it happened Don invited me to join him on the flight deck of a Vanguard sister ship from Luton to Tarbes in southern France. During the trip he told me how little had been known about Dorman, the pilot killed in the crash (pp. 1–5), among the Invicta airline crew. He was as stunned as anyone when the truth about his past came to light.

In the event, it was two days after the Paris crash before I found myself at the accident site. By then it was ringed by officials and police. Military helicopters buzzed overhead while the theory still running was that there had been a bomb on board. I gained access through this formidable security screen by the usual diplomatic skills of Anthony Terry, our peripatetic man in France and Germany. He had the toughness of the ex-military security man that he was, usefully combined with the appearance of an eminent diplomat. With assured savoir-faire, he waved some pass at the gendarmes (the right colour and shape had a magical effect) and, as an old hand, he won respectful nods from ministerial cohorts at the scene.

In less than no time we were engulfed in a posse of high officials who were inspecting the site. There will never be any point in describing the detail of that carnage. What happens when 346 people in a jumbo jet hit the ground at 497 miles per hour is best left to the imagination. Only the larger sections of the plane were recognizable – the engines lay widely scattered deep in the untouched forest far from the crash site. The plane's shallow dive had cut a huge swathe through the pine trees, leaving a scene of devastation that resembled pictures of Flanders battlefields of the First World War. The scorching fires from a full load of fuel added to the holocaust.

The faulty cargo door, the proximate cause of it all, lay crumpled in a field several kilometres towards Paris and 13,000 feet below the point on the flight path where the cabin decompression had occurred. Nearby were a few bodies, still strapped to their seats, of passengers who had sat near the door and been sucked out with it.

By this time it was already clear not only that this was the worst aircrash on record, but that the search for its probable cause was stirring profound doubts about the safety of DC-10 fleets the world over. We at the *Sunday Times* had known about the Windsor incident from the start (pp. 54–5). Yet, curiously, among the debris and ashes there had been no mention of Windsor, even among the anonymous US experts who had by this time arrived at the scene. The company's relentless public relations campaign seemed to have started even at this early stage. When we reported back to our editor, Harry Evans, in London, he was at once seized by the scale and the public implications of the disaster. He pulled out all the stops to mount an investigation that finally snowballed into a two-year operation by a team that spread itself from California to Turkey.

A first step was an attempt to reconstruct the Turkish flight on film. Before the week was out, I chartered a ten-seater Islander plane from north of London and flew it down to Gatwick to pick up a filming team. Ousted from my seat by two commercial pilots, we flew on to Paris's old Le Bourget airport – the nearest to the crash site, which by then was a totally prohibited area for over-flights. Gaining special permission to do so seemed a forlorn hope as our team squatted on the tarmac with costs clocking up by the hour. About to fly home in dismay, once again Anthony Terry came to the rescue. From some unknown French ministerial enclave he telephoned us to stand by. Within the hour, our flight clearance suddenly came over the teleprinter and the filming of a second descent to Ermenonville forest was safely in the bag.

Part of our team then set up quarters in Paris, joined by US lawyers acting for the dependants of the crash victims. According to French law, a preliminary inquiry was being held near Orly airport, to which I gained corridor access with the help of the French pilots' union. What I gleaned there suggested that evidence from the Windsor episode had still not surfaced, and the American contingent confided that the crash was now thought to have been caused by an Orly baggage-handler who had simply failed to close the cargo door properly.

The action was now moving to the United States and Paul Eddy and his team left for New York. Meanwhile, at the invitation of Turkish Airlines, I flew from London to Istanbul on the flight-deck of a surviving DC-10 – the sister ship of the lost plane. The pilots were rightly anxious to clear their dead colleagues of any blame for the crash. Inevitably, pilot error was being canvassed as a possible cause, and there were whispers that its crew should have been able to save the stricken plane. Hand-in-hand with this were suggestions that THY, which had only recently acquired its first two modern DC-10 jets, was not yet competent to run and maintain its jumbo operations. We at the *Sunday Times* already knew that a McDD liaison team in Turkey had voiced some earlier doubts on that score.

Besides hearing the airline's version of events, another key target for me was the content of the tape from the crashed plane's cockpit voice recorder. Remarkably, this had been recovered intact from the fragmented remains of the airliner and its contents promised to reveal some indisputable truths about the cause of the crash. In the event, it did. In the last few recorded minutes of the flight as the plane was plunging from 13,000 feet to earth, the sounds on the cockpit and the talk between the pilots confirmed the fatal sequence of events triggered by the loss of the cargo door. As we have seen, the resulting explosive decompression in the cabin paralysed the plane.

133

In the search for clues after an accident, the first resort is a thumbing through of the records for similarities from the past. Going back to the day of the crash, this was done within a few hours of the news,* and quickly revealed the Windsor DC-10 incident as almost certainly the true explanation of the 'carbon-copy' replay at Paris. In the next few days, it seemed strange – or perhaps not so strange – that the parallel had not yet been drawn elsewhere, least of all by the industry. Windsor clearly pointed to mechanical rather than human failure, and the talk about pilot error seemed to be mischievous, if not downright disgraceful on the evidence so far available. Later, McDD were to persist in laying the blame on the part played by the baggage-handler at Orly.

Before leaving for Istanbul, my briefing from the foreign department suggested that the Turks are a guarded race until their confidence can be fully won over, and that this could not be done in too much of a hurry. This proved to be true. Armed with my doubts about pilot error theories, and despite the usual official veto, I talked the night away with Turkish airline crews in London. We pored over DC-10 airplane manuals and made free-hand sketches of the plane's control systems. My immersion in the technical puzzle appeared to establish my credentials, at least as an attentive pupil, and when we broke up the chief pilot invited me to join him the next day on a pre-delivery test flight of a B 707 from Stansted airport.

Like many other emergent airlines, Turk Hava Yollari relied on ex-military pilots with plenty of jet-fighter handling experience behind them. Next day the Boeing was duly put through its paces in an exhilarating, if slightly unorthodox, test flight that ended with a jolly champagne party on the tarmac. The military aura of the airline extended to its disciplinary code, and I learnt that unoffical contacts with the press carried severe 'security' penalties. I was therefore surprised when, after some exchanges over the wires with Istanbul, I was offered the jump-seat on the flight deck of the next DC-10 flight back to Turkey.

After arriving there I found that my stint on the ground was not to be so easy. On behalf of the US attorneys with whom we were working, I tracked down the Istanbul lawyers who were seeking

* First credit for this should go to John Godson, an aviation author, whose good memory saved valuable time.

compensation for the Turkish claimants, in the hope that they would join in a single action to be brought in the US courts. But I sensed that they still feared that blame might fall on Turkish Airlines or its pilots. They were clearly not disposed to co-operate much at that stage, although the relationship improved as the case against McDD hardened.

The quest for the contents of the cockpit voice recorder did not promise any better. Military officials, working to their political masters, parried all requests for access to tapes and records. After a couple of weeks of fruitless interviews in ministerial offices, the sun again broke out with an invitation to lunch in the pilots' mess at Istanbul's Yesilkoy airport. Now joined by our local Turkish-speaking stringer, Metin Munir, when the convivialities were over we were beckoned to a quiet corner.

What then ensued is better left unsaid for obvious reasons, but I can affirm that no letter of the law was breached. It is enough to say that we were soon to know the content of the cockpit tape. It proved beyond doubt that the pilots of the Paris plane had faced the same sequence of events as had Captain Bryce McCormick at Windsor. His DC-10 had been less heavily loaded, and a cocktail bar had been installed at the rear end over the suspect section of the floor that collapsed under the decompression. The control lines routed under it were damaged, but not totally severed as they were to be on the Turkish plane in which the entire floor collapsed due to the weight of a full load of passengers. McCormick had some vestigial control left to save the plane, but the Turks had none.

It is worth noting that a copy of the report of the Windsor incident had reached THY in Istanbul only some days before the Paris crash. Whether any of the Paris aircrew had seen it, let alone studied it, has never been clearly established.

The voices on the Ermenonville tape – hard to identify – record the one minute and 12 seconds from the time of the decompression to impact. At first baffled by the cause, the crew found that most of their flight controls had become jammed or useless:

'Oops. Aw, Aw,' said one.
'What happened?' – Berkoz the pilot.
'The cabin blew out' – another.

'Are you sure, then?' – Berkoz again, and soon: 'Bring it up, pull her nose up.'

'I can't bring it up, she doesn't respond' – co-pilot.

'Nothing is left' – possibly the flight engineer.

'Seven thousand feet' – another voice.

(Sound of a klaxon warning that the plane is beyond the safety – never-exceed – speed.)

'Nothing is left' – unknown voice.

'Hydraulics?' – another desperate voice, probably Berkoz, calling for a check: the hydraulic systems power the flight controls. Silence seemed to tell the worst.

'We have lost it . . . oops, oops' – unknown voice.

'It looks like we're going to hit the ground!' – Berkoz, and again two seconds later he exclaimed: 'Speed!' Five seconds pass, then finally:

'Oops.'

There were no more words on the tape. In the remaining 11 seconds before impact the tape only records the unlikely background sound of three men whistling a tune in unison. It turned out to be a comic jingle from a current Turkish television commercial. On this note of fortitude, the recording ends as the plane's shallow dive cut through the tree-tops and disintegrated with near-explosive force.

The other vital tape from the flight data recorder showed that towards the end of the plane's trajectory, the two remaining wing-mounted engines had been put on full power. The effect of this, as Bryce McCormick knew when his elevator controls had failed, was to lift the nose of the plane and decrease the angle of descent. It was a last resort that saved him, but for the more heavily loaded Turkish DC-10 it came too late.

At the newspaper in London, there was no call to exploit this recorded human drama precipitately: what it had done was to confirm that Windsor had indeed been played out again at Paris, and it gave added reassurance to our team's efforts – in London, Washington and New York – to crack the whole truth as it advanced through thickets of legal obstacles, misinformation and corporate silence. It was to take two years before the alliance between plaintiff lawyers and journalists knew that they had finally breached this wall of secrecy.

If there are any doubts about Stuart Speiser's claims of uniqueness for the massive investigation into the Paris disaster, at the least it has proved to be a working model for innumerable other cover-ups in the industry. A welcome side-effect was that when it became knowledge, the charges of paranoia often thrown at sceptical and intrusive journalists became a shade less shrill.

A year before the final publication of the story, a tie-up between Thames Television in London, WNET/Channel 13 in New York and the *Sunday Times* produced a film about the Paris crash (*The Avoidable Accident*, 1975). At the publicity preview, a senior editor from the London aviation press leapt from his seat with the remark, 'This does no service to aviation safety', and indignantly strode out of the studio. I could sympathize with him to an extent – it was a brash and punchy media presentation, made before all the facts were known (but in the event they were to justify it fully). I wondered what he would have said if I had had the courage to catch him by the sleeve and ask him, 'Just how else would you do it?' Would his account come a year or more later, in the pages of an industrial press not notably given to self-criticism? Why should the public not be told the facts that were known right now?

In due course, the aviation press would have published abstracts of the official accident inquiry when it eventually appeared. This would certainly scrutinize the immediate causes of the crash and establish the link between the Windsor and the Paris flight. There was little doubt that recommendations about strengthening the cargo door mechanisms would follow and be implemented (as indeed happened).

But there the investigation would halt. The fundamental flaws in the apparatus of safety itself, as a common source of danger throughout the industry, were beyond the remit of State accident investigators. The Applegate memorandum (p. 54) would have remained in the archives unseen, and the FAA's 'gentlemen's agreement' with the industry would not have been brought to the surface to allow the US Congress to pass its severe judgment on an affair that would otherwise have remained as just another accident report.

The decline of the media

From the post-war years to the present, the media have brought the strongest challenge to the lingering paternalistic habits of wartime government. To many eyes this role of the press and television reached its zenith in the 1960s and 1970s, and it is more than a personal view that its cutting edge has become blunted over the last decade.

It is enough to say that investigative journalism as a slogan – once carried on the banner of an inquiring press – has become a pejorative term today. With some honourable exceptions, the new order demands up-beat stories and puts slogging detective-work into the seamier side of the news at a discount. The consequences for air safety are not good, in so far as it is the media's job to keep the consumer informed. Current anxieties such as congested airspace and the danger of collision, the lowering of safety standards following the deregulation of US airlines, or the continuing number of fatal aircraft fires, tend to be reported in fairly general terms without any deeper penetration behind the flow of publicity handouts and press releases.

The assumed right to official silence has permeated everyday affairs, and there must be many other journalists who now lift the telephone with, as it were, a threatening sense of exposing their heads above the parapet into heavier legal fire than before. Little wonder that the less adventurous keep their heads down.

The deployment of the weapon of secrecy is not confined to routine affairs, but reaches to the highest levels. An illuminating report in the London *Times* by Michael Binyon of the Euro-summit meeting at Madrid in June 1989, tells how – unlike other countries – British press briefings are always given on an unattributable basis and no television cameras are permitted. He notes 'the smack of contempt and cynicism with which most British civil servants regard Fleet Street' that accompany such proceedings. The Germans' briefings, he says, could not be more different, in which cabinet members speak freely and expect to be quoted, while the French offer on-the-record comment by senior diplomats as a matter of course. The comparison provides a litmus test that is hard to gainsay.

It is becoming more difficult to deliver the news. Less is drawn out into the national media, and news reports tend to be shorn of specifics or sanitized to satisfy the lawyers. A sample from a list of events that have attracted little notice seem to confirm this. Some are facts from the technical press and others must be treated as anecdotal until the specifics are verified, although there is no reason to doubt their origins or their intrinsic truth.

- A 1988 international seminar on flight safety was told, for instance, that since 1959 the world's airlines have wrecked 337 jet hulls – surely a fact to conjure with, but not heard in public comment about the airline industry.
- The Portuguese island of Madeira is a much-favoured winter sunspot for northern Europeans. Its Funchal airport also has a dauntingly short runway, bounded by a cliff-edge above the seashore. In 1987 the pilot of a British Boeing 757 charter plane, hassled by the imperative of an accurate touchdown, made a very heavy landing and smashed its undercarriage in the process, as testified by the reported cost of repair at no less than $50 million. The Portuguese authorities were riled when the British subsequently removed both the plane's safety recorders which, they claimed, was done in defiance of international regulations. Now there are rumours that the airliner may be impounded at Funchal, where it still lies. But there is a second act to the story. Not long after this happened, another British charter plane – this time a B 767 – attempted an approach to the same airport, but the pilot thought better of it. He slammed on full power and took off over the cliff's edge, to disappear 'like a fighter from the deck of an aircraft carrier'. Mystified, the Portuguese telephoned the chief pilot of the airline in Britain, to ask if they had a pilot who had made this notably dashing manoeuvre. 'Yes,' answered the chief pilot, 'it was me, and I'm going to cancel the route.'
- Then there is a 1988 report of a British Airtours TriStar (L-1011) whose pilot had a nasty moment when he brought the empty plane in to land at the little local airport at Cambridge, England, for engineering repairs. With a runway about half the length of that at London Heathrow, the stopping distance is marginal even for an unladen jumbo. Following usual practice after touchdown, the pilot simultaneously applied his brakes and reverse engine thrust, but the brakes failed to work. With an insight given to few, he pulled the throttles back to cancel reverse thrust and tried

139

his brakes again – and this time they worked. Reverse thrust was resumed and the plane stopped safely on the tarmac. There is a hi-tech computer that controls braking power, and engineers are now testing the theory that when engine vibration passes through a certain range (11 to 14 Hz), the computer goes on the blink. Whatever minor fix may be needed to cure this unforeseen flaw, one wonders what might have happened to a fully laden plane landing at one of the many other airports with marginal stopping distances. Perhaps the rogue computer was unique. Yet it happened.

● On the positive side, it is equally strange that some of the good news for air passengers seldom comes to the surface. It is more than a decade since the airlines were persuaded to adopt a magic box that warns the crew when a plane is in danger of flying into mountains or comes too close to the ground. The versatility of the ground proximity warning system (GPWS) is more than this suggests, and in its many diverse roles it has undoubtedly saved many lives. 'Flight-into-terrain', a common cause of accidents in the past, has become markedly less so. As a Boeing safety man found recently, 'nowhere in the safety statistics has one change so clearly showed up.' If it is assumed, for instance, that the proportion of such fatal accidents had remained unchanged over the last ten years because GPWS had not been in use, it is possible to argue that its advent has saved nearly a thousand lives. (But being based on a supposition the figure can only be a rough guide.)

But even this happy outcome has its darker side. Crashes into high ground had continued to cause needless deaths until the authorities were finally persuaded of the crying need for an on-board device to warn pilots of approaching danger. The rising death toll from such crashes in the 1960s and early 1970s culminated in a disaster that speeded the advent of GPWS: on the morning of 1 December 1974 all 92 people on board a US TWA flight were killed when their Boeing 727 crashed into mountains near Washington DC. Once again it revealed the lethal consequences of secrecy and delay.

The airliner collided with a mountain 44 miles from Washington's Dulles International airport as it was making a difficult approach through bad weather. Like all avoidable collisions with the ground, the primary cause was, of course, a navigational error. On that day it happened because there had been a muddle about who was responsible for ensuring that flights maintained the minimum safety altitudes laid down for the area – was it the pilot or the airport radar controllers? It seemed incredible that such an obvious danger should be tolerated, but that was not all. Less than two months earlier, an aircraft of another airline, while flying an identical approach to Dulles, had nearly flown into the mountains as a result of the same confusions. When its crew dutifully reported the incident to management, the airline issued a warning notice to their own flight crews but it failed to pass it on to the authorities or to other airlines. The identity of this flight was concealed because, it can be assumed, the airline feared adverse publicity – a secret that led to appalling consequences.

The Safety Board's report of the TWA crash blamed the Federal Aviation Administration for its failure to resolve a long-standing confusion between air traffic controllers and pilots which, it said, the FAA had known about 'for several years'. As to the TWA inbound flight over the mountains, it found that the airport controllers had wrongly given it clearance to approach while it was 44 miles from the airport on an unpublished route and without clarifying the minimum safety heights to be flown: the pilots, the report said, should have double-checked but they had relied on the controller's instructions. More importantly, the shock of another fatal crash into the ground – this time, as it happened, on the doorstep of the FAA's headquarters in Washington – at last demonstrated the need for an answer to one of the oldest hazards of aviation, that of flying into high ground unawares.

By December 1975 the FAA had made GPWS a mandatory fitting on all large US airliners and other world fleets soon followed suit. It also sorted out the respective duties of pilots and controllers in observing minimum safety heights.

A lot of work went into the concept of GPWS and its subsequent history shows what a large dividend can be earned from a single piece of safety equipment. But it has to be said that before its use became mandatory, the cost of fitting this sophisticated equipment to their fleets raised some strong opposition from sections of the airline industry.

● A light-hearted item comes from the memory of an airport fire expert about a non-fatal crash that happened on the runway of an international airport some years back. It was served by a notably dozy crew of the fire and rescue services who were relaxing over tea in the fire station when a giant wheel from the main undercarriage of a Boeing 707 smashed through the doors. The fate of their teapot goes unrecorded, but outside lay the B 707 with a badly scraped belly. Luckily for the passengers in it and the fire crew, it did not catch fire. The teller of the story is confident that, at least at the best of airports, fire teams nowadays can meet their requirement to reach airport crashes within three minutes. It is a comforting thought in the knowledge that 80 per cent of all air accidents happen within 1,000 metres of an airport.

Some of these random anecdotes, and many others like them, may find their way through official safety machinery to be purified and finally appear in print. Others may not. Apart from the GPWS history, none appears to have filtered through to media watchdogs.

11 | The public's right to know

The motives behind most of the diverse manifestations of secrecy that have so far been reviewed here generally speak for themselves. One important motive not to be overlooked is that often politicians wish to quell rumours not because they might scare the public unduly, but because they are frightened of the ammunition they will provide to their opponents.

Chernobyl

The implications of Chernobyl must have brought fearful consternation to the airlines flying through airspace threatened by the fall-out in the hours and days following the explosion, as meteorologists tried to predict the radiation cloud's erratic weatherborne course across Europe. (In this they were not notably successful – first assurances that the cloud was dispersed over less populated areas conflicted with a reading of the synoptic weather charts showing a change of wind that threatened to bring it back across northern Britain, as indeed it did.)

Air travellers were kept even more in the dark. The deluge of assurances about the safety of the international nuclear industry had served to shelve the prospect of unthinkable disaster not only from the public mind but also, it soon became clear, from the vision of the officials responsible for civil defence. Whatever contingency plans they were working to, the hiatus of silence that followed the catastrophe indicated that the authorities had been caught on the hop. Fear of public panic may have been one reason why air travellers were given not a word of official advice about the possible hazard to impending flights. A more likely reason is that nobody had an answer ready. Or again, that official conditioned reflex to sudden and troublesome events came into play – one that urges that at all costs, a media clampdown must be the order of the day until a considered policy brief can be fed to their press offices.

143

Questions of flight safety became a taboo topic overnight. Airlines and officials remained either sanguine or noncommittal, and the only news of the effect of the Chernobyl emissions upon flight schedules tended to be anecdotal. Others must have had friends or relations like the man who, returning from Singapore to the UK just as the news of Chernobyl had broken, had his flight diverted en route to a westerly airport and finally arrived, still mystified, in London a day late. The passengers on his plane had been told nothing about the reasons for the diversion and delay, least of all the attempt to circumvent radiation hazards.

In the midst of the Chernobyl scare another celebrated traveller, the best-selling British novelist Tom Sharpe, was returning from Switzerland to London. As he looked out of the window of his plane over the snow at Geneva airport, he watched a LOT (Polish Airlines) Tupolev 134 airliner inbound from Poland land and taxi past. 'It didn't surprise me that it looked pretty filthy,' he remarked, 'but I did wonder how much of the dirt was radioactive and if anybody checked it out.' He was not reassured by the acute observation that the wind that morning had changed from a north-easterly to a south-westerly – the wind that brought the danger back across Europe. Nor were his apprehensions relieved by any assurances from the airline either during or after the flight. The passengers were allowed to disperse at London without further enlightenment or any mention of the need or otherwise for bodily radiation checks.

Months later during the run-up to the 1987 General Election in Britain, at a gathering of aviation professionals, it was confided by one present that on the day of the Chernobyl disaster he had learnt that there had been 'sheer panic among the airlines at Heathrow'. Incoming planes were washed down as airline managers sought specialist advice. He had been told that, in the event, radiation levels on the planes were found to have been unexpectedly low. They turned out to be less than those sometimes detected on Concorde after its high-level transatlantic flights, where ambient radiation is higher than it is at normal cruising levels. Concorde carries constant monitoring equipment not only as an airline safety measure (the crews are, of course, most at risk), but also for aerospace research purposes. No 'red alert' has ever been recorded and the results to date are said to confirm that there is no measurable risk to passengers.

144

The comment from Heathrow about the reaction of the airlines was followed by an appeal to the rest of the gathering to keep these remarks within the circle on the grounds that no one would want the politicians and the media to exploit them (the fact that he was rather more specific about those who he felt might misuse it need not trouble us here). This apparently won tacit approval from his audience. While it might be thought legitimate to report this conversation without naming names, in the ordinary course of affairs it would be rather distasteful to do so even without attributions. But the rider he added about the supposed *political* implications – rather than anything to do with air safety – did, I am afraid, remove any such inhibitions. I could not resist the conclusion that here was a scenario – admittedly founded on hearsay and as such possibly apocryphal, although the speaker's sources of credentials were hard to doubt – which must have been first and foremost of profound importance to passengers and aircrew whose bodily safety was at risk at that moment, while the health of the airline managers and the authorities concerned clearly were not. Who had the greater need to know?

Some time passed before official advice about the safety of specific routes began to filter through to the public. Undoubtedly there must have been genuine reasons for the delay while agonized experts and administrators tried to assess the degree of danger to the public, but intelligence from the airports revealed that the presumption in favour of not upsetting the trade was also playing its part. Added to this, apparently, came the droll reluctance to spill the beans in case it gave succour to the anti-nuclear lobby, if not to the loony Left. If this was indeed a motive, it hardly served the interests of bewildered travellers and those who had relatives flying around Europe at the time.

Politically, it revealed a sorry gap in contingency plans for the civilian front, although just how far the authorities were taken short by the event is another story. At the time, news centred mainly on a party of students returning from Moscow whose relatives had prompted the media. If it is known how many other people were thought to have been at a potential risk from radiation and how many were located and screened, this has not been revealed.

As the true story of the airlines' reaction to Chernobyl has not been given to us, we are left to draw what conclusions we can from these hearsay fragments. Those who know more, and who are disposed to complain about the irresponsibility of disclosure, have only themselves to blame. What we can say with more certainty is that, when the chips are down and folk are risking physical harm, the first instinct of those in the know is not to tell. More frankness could do us all a bit of good. Consider, for instance, the admission of British Nuclear Fuels that 'public pressure is the only factor that accelerated the pace of safety controls' in that industry.

It was that pressure that finally brought to light, decades after the event, the near-catastrophe at the British nuclear reactor at Windscale in October 1957. Its manager, Ronald Guasden, the only man to have tackled a Chernobyl-like fire before, describes the unforeseen event and the failure to extinguish it with the hopelessly inadequate safety systems then available. Attempts to blow it out with fans and then to stifle it with gas were useless. Although he had no idea of the probable effects, he resorted to deluging the atomic piles with water. Luckily for humanity, it worked. Today he concedes that he then had no inkling that he and his staff were on the verge of the historic near-catastrophe 'that was to unnerve the nation'. But we were not to be unnerved until the aftermath of Chernobyl forced a repentent industry to end its thirty-year-long cover-up. When the facts at last became known, safety standards in the industry soared in step with the political concern fired by the public clamour.

Another question remains. Behind the scenes, was it the growing recognition of the scale of the potential disaster in 1957 that led to its successful cover-up until Chernobyl persuaded those in the know that it was time to tell? Or was secrecy just an official reflex action? It is now no longer a secret that Windscale has a military role that would justify silence about the 1957 event, yet Chernobyl broke down that wall of secrecy. The transition from the cold war may provide some explanation for this U-turn in security policy, yet the suspicion remains that habit was the strongest motive for the years of silence.

Confidentiality and the public's right to know

A second glimpse of official caution inspired by curious motives came in the aftermath of the Boeing 737 British Airtours fire disaster that caused 55 deaths at Manchester International Airport in August 1985 (see pp. 188–93). After much public criticism about the difficulty of evacuating passengers from the seating configuration used on that model of the B 737, the Civil Aviation Authority set up a programme of simulated evacuation tests at Cranfield research centre in Bedfordshire, England. But it became apparent that this venture and its venue were to remain confidential. Some of those interested in following such proceedings, however, got wind of the coming event from a colleague, and although they accepted that the official doors were closed, they fell to taunting him about other means of gaining entry. If they just happened to arrive on the appointed day, perhaps behind tinted glasses and a false beard, might they not penetrate the CAA security men? The lips of their informant smiled but remained sealed.

Whether the reason for these clandestine tests came under the heading of commercial confidentiality or some other bureaucratic inhibition, an unwelcome witness could hardly pose a dastardly threat to national security. Presumably the tests cost public money, but that is not an argument that opens doors. Any challenge to this back-room policy brings a response that is thoroughly familiar to journalists. The uninformed can get the wrong end of the stick, they are told, and they might leak it to the media. Ill-informed and precipitate judgments, so the theory runs, can distract the public and the politicians from the value of a proper professional analysis when it becomes available in their own good time. The rationale is clear and there is much truth in it, as we have seen from the mistakes made by newspapers. Battle lines are drawn along the contested frontiers between the claims of confidentiality and the public right to know.

The benefits of each must balance. The place for argument comes at the point where the two graphs cross over. The media can abuse its role by adorning the truth, or deservedly exposing a culprit. The workings of government and industry can suffer from naive and inaccurate leaks, or be cleansed by the ever-present risk of being put to public judgment. But time and time again, the media – however crude, intrusive or misleading their antics often are – still earn the title of watchdog-in-chief. Those in authority who are tempted to abuse their powers are surely less likely to do so since the presidential wrong-doing exposed by the two Watergate reporters, a scandal that has provided a refreshing model for a number of less celebrated revelations on both sides of the Atlantic. Yet there are still those who seem to doubt the benefits of an inquiring press, or who at least demur at the methods it employs. One prominent British commentator, Paul Johnson, recently compared the BBC's use of investigative reporters ('a low form of journalistic life') with share ramps in the City.

Whatever depths of iniquity may be attributed to our masters, it is certain that their first concern has been to protect official confidences. Authority abhors a leak, but in its eagerness to maintain security about its affairs, it seems bent on shooting the messenger – the media – rather than disciplining its own whistle-blowing servants. The corporate or official image must be protected. The point was better made in 1982 by Archdeacon Brookes who, in a Shakespearean allusion from *The Tempest*, compared the hatred of those in authority for the press and television with the monster Caliban's rage at seeing his own reflection in the mirror. The fact that popular papers sometimes resort to a distorting mirror does not detract too greatly from the validity of his insight.

As the record of aircrashes shows, in the field of air safety some of the worst effects of secrecy stem from delay. When the final report of a crash is published, it too frequently shows that the cause has been known for a long time by some but not disseminated to the many. In recognition of this and to the lasting credit of British Airways, the airline publishes full details of both minor and major operational mishaps in its monthly *Air Safety Review*. In the name of safety, BA is prepared to stand the unfair consequences of baring its soul in print to its competitors, and as a result its mishaps also reaps more than their fair share of attention from the media.

When the British Concorde, for instance, suffered an agonizing period of failures in all its four hydraulic power systems, the anti-Concorde faction read about it with glee in the world's press. Those who had the temerity to approach Air France about similar problems on its Concorde fleet (or indeed any safety problems) were met with monosyllabic replies.

Although the BA *Air Safety Review* was ostensibly confidential to pilots and professionals, it was in practice readily leaked to those with a genuine interest in safety and to a limited number of trusted journalists in the field. Then came two articles in a leading British national newspaper which showed themselves to be based either on a gross misunderstanding or a delicate distortion of data contained in

148

the *ASR*. Before publication a decision to withdraw them for these reasons was over-ruled by hungry but misinformed news chiefs. One of these stories fixed upon a relatively minor misbehaviour of the nose wheel on one of BA's Concordes and implied that its passengers had been put into far greater danger than in fact was known to be true. The second article floated a rumour that a recurrent design flaw in the airline's Boeing 747s was about to ground its entire fleet of jumbos.

This brought angry denials from the airline's chief, Lord King, and soon after the two stories had appeared the Reviews were put into severely restricted circulation. Current and back numbers were withdrawn even from the shelves of the Civil Aviation Authority's London library. So it could be seen that some of the loudest critics of Britain's closed and secret industrial culture had shot themselves in the foot, through the careless mishandling of material that had been entrusted to them and which they did not understand. The episode invites a rider that appears to be apt here. Specialist writers know that they have to live with the well established delusion that a good journalist can write about anything. Jacks-of-all-trades may be a godsend to news editors harried by late stories that break as deadlines approach, and in the short term they may have to fill the breach.

But readers will soon know that the reportage of subjects lying in their own fields of knowledge is ill-informed, if not downright piffle. It's a common complaint. Their incredulity is then likely to spread across the other pages in the newspaper and, at worst, the newspaper will lose those readers. If a sustained circulation depends upon the reliability of a newspaper's reporting, to that extent the idea of the polymath reporter must take a severe kick in the vitals.

The long-term consequences are more widespread. Irresponsible or uninformed reporting strengthens the case against media interference and encourages officials to bolt their doors more firmly against future appeal for information and help, from both their trusted and less trusted inquirers. The public will also suffer, if it is conceded that the media can now and then lend a hand in speeding up the progress towards safer flying – and taking the bad and the good together, the record shows that this is not a wholly vainglorious claim. For the health of the quality papers and television programmes at least, if reliable reporting is seen as a long-term selling point, then the eternal friction between those who know and those who wish to know does need to be lubricated by good faith on both sides of the fence. Parity of knowledge of the subject matter wins smiles and confidence. This is not to say, of course, that poor or unscrupulous reporting should in any way justify official secrecy: but rather that in order to score a goal the rules of the game must be obeyed, or players must risk being sent off the field in disgrace.

Returning to the examples of secrecy as a threat to safety, they may give a sight of the incubus that hinders the truth from emerging at the right time and to those with the first right to know it. These may be the general reader or the 'buyer of last resort' in the shape of the air traveller worried about the safety of his person or his family. The power of institutionalized secrecy leans on the need to keep him in a state of ignorance, if not to disinform him. Only in those rare climates in which notions about the sovereignty of the consumer are something more than convenient political cant, is he likely to be given open access to the material facts about the product on sale. Elsewhere, what can be done to make him a wiser and more discriminating customer?

The surest cure for secrecy as a chronic distemper of British institutional life remains the *prospect* of being found out – the chance that there may be a public eye looking through the key-hole, an eavesdropper or a receiver of leaks at the ready. The answer to those indignant moralists who throw up their hands at the whistle-blowers is surely that handsome is as handsome does. Sneaky methods beget sneaky people on both sides of the closed doors. And those with a distaste for the sort of investigative journalists who call for more open government can relish the irony that, if this ever came about, that tribe of inquisitive gentlemen could have worked itself out of a job.

Part 3

Into the twenty-first century

12 | Across the oceans

'If you need an accident to know you have a problem, then you are part of the problem.'

Chuck Miller, ex-Chairman, US
National Transportation Safety Board

We have now seen the web of secrecy which entangles the whole airline industry, from the manufacturers to the carriers, from governments and regulatory authorities around the world to the media. In Part 3 we will try to break through this web and discover how the industry is tackling the three crucial issues of the next decade: the doubts surrounding twin-jet engines, the terrible risks of fire on planes, and the dangers of over-crowded skies.

I hope it is true that an Irish joke may be re-told without offence if, like the following old story, it first came from the lips of one of their kind. Whether or not its shelf-life has expired, it launches us into the realms of flight without engines and, for once, without any terrible after-effects. Two of my Irish friend's compatriots, he said, were flying home across the Atlantic when the captain of their four-engined jumbo announced that they would be a few minutes late because one of the engines had failed. Shortly after came further regrets that the delay would be rather longer as a second engine had quit. Minutes later, of course, they heard the captain's reassuring voice announce that they would be an hour late as a third engine had conked out. One worried traveller turned to his mate and, with impeccable logic, whispered: 'If the fourth engine goes, we'll be up here for ever?'

As no big passenger jet has yet made a fatal plunge into the icy waters of the North Atlantic, such hoary stories can still be retold in moderately good taste. Just how long this will remain so is a much more sobering thought, because planes with only two engines are already beginning to replace the bigger three- and four-engined jets that up to now had to be used on long-distance ocean routes and for flights over remote territory. But the claim that the new breed of long-range twinjets are as trustworthy as the three- or four-engined planes has plunged the industry into a bitter controversy. Too many narrow shaves caused by twinjet engine failures have put pilots, passengers and airlines on guard.

A stop-press item appeared in a London paper on Saturday 16 January 1988 about a transatlantic jet that had to fly 400 miles over the ocean on only one engine after its other engine had failed. It had taken off from Gatwick the day before bound for St Louis, but was forced to make an emergency landing at Goose Bay in Labrador. The US news services did not carry the item and the aviation press later gave it only a cursory mention. The profound significance of this apparently minor event escaped the notice it deserved. For it represented the first report of an engine failure leading to emergency landing done in anger, so to speak, on the North Atlantic routes flown by the new twinjet airliners. Six months before another twinjet lost power from *both* its engines and swooped down to 500 feet above the Pacific Ocean before the crew managed to restart the jets and avert disaster.

These two items come from a formidable list of recent twinjet scares that have not helped the industry's claim that the new breed are as safe as the old. There is no doubt that, in theory, twinjet flights on these routes *could* match the safety standards obeyed by the bigger planes, but this has yet to be proved in practice. Meanwhile the world's regulatory authorities have set up tough new requirements for long-range twinjets that must be met before they can fly the routes. In outline, they come in two parts: the planes must at all times be able to reach an airfield if an engine fails, and they must carry extra back-up equipment to ensure that they can do so. In fact, the two rules are inter-related because the more back-up gear they carry, the further they may fly from an airfield. The argument that is now breaking around the ears of the industry is just how realistic these safeguards will prove to be in practice. The line-up shows the pilots, the customers and the safety regulators in the camp of the doubters and the cautious. They are opposed by an industry that seeks the benefits of much cheaper twinjet flying costs. Safety is the football that lies in the middle ground between them.

154

The reliability of twin-jet engines

Common sense seems to argue that it must be safer to fly with more than two engines, and many still firmly hold to this view. But the plane and engine manufacturers as well as quite a number of airlines insist that the higher reliability of the new generation of airlines now matches that of the planes of yesteryear. Since the early 1980s the controversy has raged as the regulatory authorities have tried to hammer out a compromise that is the least limiting and costly to the airlines. Unless the old rules that governed three- and four-engined long-haul planes are relaxed, operators must bend their twinjet routes to keep within the safety-radius of airfields en route. On the North Atlantic, this takes them away from the direct routes on a wider arc to the north closer to Iceland and Greenland.

For all those who may find themselves faced with the prospect of a long-distance flight with only two engines to support them – and there are now some 1,600 flights a week worldwide flying these routes and others bidding for the freedom to do so – it is worthwhile to follow the argument with the aid of some recent facts. Nobody would deny that the new breed of long-range hi-tech twinjets are magnificent aeroplanes. At present the main contenders in the field are the two wide-bodied planes – the European Airbus assembled in France, and its rival the Boeing 767 – and the slimmer Boeing 757. It is certain that they are capable of proving themselves to be up to the job – given time and experience. But it is a reassurance to see signs that their final approval is likely to be given in the long run by none other than yourselves – the fare-paying passengers.

If the 142 passengers who found themselves stranded at Goose Bay were irked by the inconvenience, they should count themselves fortunate for two reasons: their TWA Boeing 767 flight had duly observed the safety rules and kept within its 'single-engine' range of a diversion airfield, and the fault affected only one engine and not two – an event that is far from unique, as we shall see. Strangely enough, the episode got far less attention than another departure from Gatwick only eighteen days later, when a Continental Airlines Boeing 747 lost power from one of its four engines soon after take-off. Breathless press reports told of its captain jettisoning fuel over the English Channel before returning safely to Gatwick. It is true that the crew had some nasty moments handling a heavily laden

155

jumbo in a gusty crosswind after losing an engine at the most crucial moment during the take-off, but this is not an uncommon event. Unknown factors apart, it is odd that the loss of one engine out of four close to an international airport should stir the adrenalin to more effect than the loss of one out of two in mid-Atlantic.

Before entering the scene of planes battling with the elements in the more chillingly hostile parts of the globe, it seems wise to ask what the pilots think about it – the men at the sharp end of the matter. Although the distance between the two opposing camps has shortened, the pilots' doubts are still heard loud and clear above the assertions and counter-assertions in the debate about safety and reliability. The anxiety of pilots was broadly reflected in the chairman's report (1986–7) of the Air Safety Group in Britain, an advisory body of aviation experts headed by parliamentarians. The chairman, an ex-airline pilot himself, declared: 'I am firmly convinced that the majority of North Atlantic passengers, particularly the package holiday passengers, are not aware that they may well be transported across the North Atlantic in a twin-engined aircraft. I have a great deal of misgivings in the matter.' He added that if he was boarding a transatlantic flight as a passenger and found it had only two engines, 'I would walk off it'.

When safety is the first concern, the experts seem to agree. In the United States, Air Force advisers came to a similar conclusion about a replacement for the elderly presidential plane, Air Force One. According to *Time* magazine (22 January 1990), they refused to look at any plane 'with only two motors', such as the new long-range Boeing 767 twinjet. Consequently, President Bush is to take delivery of a four-engined Boeing 747 jumbo in September 1990.

Any attempt to summarize the technical arguments is bound to be selective, but some of the key points are clear enough. The plane and engine manufacturers, including Airbus Industrie, all cite much the same high order of reliability for the twinjets. One major engine manufacturer, Pratt and Whitney, is reported to claim that the projected rate of inflight shutdowns is one in every sixteen operational years, while the International Federation of Airline Pilots' Associations (IFALPA) asserts that shutdowns will happen once in every four months. A British member of that Federation, Steve Last, believes that there is 'a devil of a lot of complacency in the industry . . . there is a lot of twaddle spoken about engine reliability statistics'. He has even produced figures that, he claims, show that the (two) piston engines on the old prewar DC-3 Dakota had a lower shutdown rate than today's big fan-jet engines.

While the two sides bandy figures about, the facts are that between 1982 and 1989 at least 17 passenger airliners have either lost all their engines in flight or have been reduced to a single engine. All but three of these involved the new twinjets, and four of these lost *both* engines in flight.

By September 1989, Rolls-Royce had won the approval of the UK Civil Aviation Authority for 180 minutes single-engine operation of its RB211–535E4 and C turbofan engines, thus allowing planes such as the Boeing 757 to fly routes that are as much as 1,400 miles from the nearest airport – in practice, this means almost any route in the world. The extension will soon be given to RR engine models to be used on the B 767, and American approval for both is expected to follow soon. Planes fitted with these engines would, of course, have to comply with all the other requirements for extended-range flight, but it is notable that the reliability of the engine itself was greeted by the manufacturers as a breakthrough for the twinjets.

But it is now conceded on all sides that the reliability of the engines is far from being the only measure of twinjet safety. The chances of breakdowns in equally vital systems provide less back-up than there is available in the more generously endowed three- and four-engined big transports. The buzz-word for this is 'systems redundancy', and the smaller basic twinjet models naturally have less of it than the multiple and independent systems built into, say, the Boeing 747 jumbo. More will be said about these later, but those that are most vulnerable on the twinjets group themselves under the headings of electrical systems, cargo fire suppression devices, the auxiliary power unit or APU, hydraulic systems that power essential controls, and adequate cooling for the mass of electronics (or avionics) on which flight management and navigation depend.

Experience already shows that a fault in any of these groups, whether or not compounded by an engine failure or shutdown, can spell an emergency on a twinjet flight. Its bigger brothers can take a similar fault in their stride by switching to one of their multiple back-up systems. Aware of this, the regulatory authorities are now proposing either higher proven reliability in the systems or additions to them. While the tug-of-war between the industry and its critics continues, the safety argument now centres mainly on whether or not these mandatory safeguards will cut too deeply into profits on the one hand, or be tough enough to ensure safety on the other.

Very recently there have been signs that the industry's early enthusiasms have been dampened for what they saw to be cheaper two-engine, two-crew long-distance routes. It is not only a matter of cost benefits. As one airline admitted, there is 'a growing public awareness' of the internal controversy. In other words, it has leaked out and led to distinct customer resistance. British Airways' Lord King, for instance, has firmly rejected the idea for the time being. Reportedly, he believes not only that more experience is needed before the twinjets are cleared for long-distance passenger flights, but also that 'passengers could be put off by the thought of relying on one engine should the other have to be shut down when the aircraft is up to two hours away from the nearest airport'. Lufthansa's response to the idea has been an uncompromising 'never'.

For once it seems that the ball has been passed to the customer. To him, it *sounds* less safe to fly on two engines, given the option of flying with three or four. If he is to be persuaded otherwise, he will expect to know the reasoning that has led the regulatory authorities to put the twinjets in the same safety league as the big jets. And much of this turns upon whether the kind of safeguards they are now proposing really meet the facts of life that have prompted them. As so often, the safety rules that govern long-distance flights start from a base-line set well back in the past.

The new safeguards

We are told, then, that a new package of safeguards applied to the twinjets flying on long-distance routes will ensure that they are as safe as the three- or four-engined types they are to replace. Before hearing the arguments for and against this proposition – and the practical ifs and buts – there are two points that are worth outlining at this stage.

First, it is obvious enough that the failure of one engine on a twinjet puts a premium on the reliability of the remaining one. The loss of an engine is therefore classed as a full-scale emergency and this demands an immediate diversion to the nearest airport – hence the careful calculations about its single-engine endurance. There is also the chance, as the record of such incidents warns us, that the cause of one engine failing may spread to the remaining good engine. Fuel contamination or shortage are examples. It is therefore possible to argue that the reliability of the remaining engine becomes more suspect than it was before the failure of the first. And as will be seen later, single-engine flight brings many other hazards into play – for instance, more fuel is needed to travel a given distance than would be used in normal two-engined flight, so that the distance to a landing becomes critical.

Second, the list of twinjets that have come to grief recently is quite formidable. It must be borne in mind that many of these were not flying transoceanic routes and so were not subject to the new safeguards that now apply to them. None the less, since so much of the case for twinjet safety rests on their inherent reliability as a breed, as compared to planes with more than two engines, the incidents on non-oceanic routes cannot be discounted from the overall question of twinjet safety. But the differing operating environments that are met on short- and long-haul flights have been recognized since the early days.

The safety rules governing long-distance routes that traverse oceans and inhospitable terrain reach back to the 1950s, and they were intended to apply to all airliners irrespective of the number of their engines. These routes now come under the ugly title of Extended Range Operations or EROPS (or just ER for short) and the acronym is also borrowed to describe aircraft types that meet the required standards for such routes.

The cardinal rule was that planes had to be able to reach a suitable airfield if one engine should fail. This was no problem for three- or four-jet planes and it did not restrict their ability to fly the shortest routes across the Atlantic and elsewhere. The advent of the twinjets meant that they had at all times to be able to reach an airfield on their one remaining engine. And as we shall see, the consequences are considerable and a whole heap of hazards pile up on a pilot when he is forced to fly to safety after losing 50 per cent of his engine power and back-up systems.

In fact, the evolution of the rules for ER flight was more complicated than need be told: routes had to be adjusted so that at any point, and assuming an engine failure, planes could still reach an airport within a given time limit. The US Federal Aviation

Administration set a basic limit of 60 minutes, and may extend it to even 180 minutes according to the reliability of any candidate airline and the standard of extra equipment fitted to its aircraft. In Britain, the Civil Aviation Authority is taking a much more cautious approach to the 180-minute proposition, as are its European counterparts. Some confusion is added by an advisory standard of 90 minutes for ER twinjets set by the International Civil Aviation Organization, and by recommending its own standards for ER-equipped aircraft, but the detail need not concern us.

The 60-minute limit was restrictive enough to require flight routes to be lengthened so that they stayed within their time-radius of emergency landing grounds – on the North Atlantic, for instance, they would be bent to the north closer to bases in Iceland, Greenland, Labrador and Newfoundland. But under airline pressure, the authorities are willing to relax the time-limitation on the *ad hoc* basis described. Given the full complement of added extras, many regard 120 minutes as a safe enough 'single-engine' time-radius for twinjets to reach a emergency landing ground, and this would allow them to fly the normal direct routes across the Atlantic. This is clearly the ambition of the airlines, and Boeing now supplies ER-equipped models direct from the production line so that the gateways to the shorter routes would appear to be opening up. But it will become evident that there are other constraints on the freedom to use them.

Additional safety equipment is only one reason for the new elasticity of the 60-minute rule: different states, as well as manufacturers, propound or advance the reasonableness of longer safety times. France and Airbus Industrie, for example, adopt the 90-minute ICAO standard. But as the plane-makers and a number of airlines are pushing forward towards these wider time-margins in order to cut operating costs, the pilots are cautiously applying the brakes.

Opposition from pilots

Not forgetting that much has happened in the last few years to draw the two sides closer together, the pilots' misgivings were fully aired at a meeting of the International Federation of Air Pilots' Associations (IFALPA) held in London in 1983 to which I was invited. To an extent, the professional concerns aired at that meeting stimulated many of the safeguards that are now becoming part of the twinjet regulations, but these still represent a compromise that has left the aviators with some very real anxieties.

At that time Boeing was busily canvassing a relaxation of the US 60-minute rule and claimed that the new planes could meet the current ICAO 90-minute limit for twinjets with safety in the face of mounting opposition from IFALPA. The pilots alleged that the aviation press had 'relied heavily on aerospace companies' advertising' and had simultaneously printed 'planted' articles stressing the long-range safety of the twinjets. They suspected that the industry was pressing its case behind closed doors in the ICAO building in Montreal. They then produced a formidable list of dangers which they claimed were being bypassed.

First, IFALPA argued that bending the routes to the north could not be commercially viable and would effectively add an hour to the flight time across the Atlantic. Even under the tighter 60-minute rule, twin-engine flights would be hazardous, and to increase this to 90 minutes would enlarge the risks by routing far further from landfalls. Either way, they feared, the routes could only be flown by twinjets by ignoring a host of operational hazards that could follow an engine failure in mid-Atlantic. A list of these perils read together admittedly present the 'worst-case' possibilities that could face an aircrew, but they are none the less real for that. Although some of these fears have been allayed by the new rules, they are worth reciting as a vivid description of the practical risks of flying an airliner on one engine.

● The speed drops from about 520 to 460 m.p.h., not only because the plane has lost half of its available power by the loss of an engine, but also because it can no longer maintain its normal cruising altitude around 40,000 feet and must descend to its single-engine 'ceiling' of about 23,000 feet, where the denser air offers more resistance. Added to this is the plane's tendency to fly crab-wise under asymmetric power from one engine: although this can be 'trimmed out' within limits, the remaining drag still leaves it in a less efficient configuration. The lower speed, of course, stretches the time needed to reach a safe landfall.
● Normal cruising levels are 'above the weather'. Although there can be strong winds at high altitude, these are known from pre-

161

flight forecasts so that time and fuel allowances can be calculated in advance. A descent to 23,000 feet can plunge into unknown winds and turbulent weather that slow the groundspeed of the plane even further – for example, a change from a tailwind to a headwind could knock off up to another 100 m.p.h. or even more. Pilots complain that weather data for the lower levels or long routes is not available either before flight or from en route radio services (and Atlantic weather ships have recently been withdrawn). Calculating the flight time to safety is therefore impossible until the emergency strikes, and by that time the weather beneath may be such as to put landfall out of reach.

● Each engine is a separate source for the electrical, hydraulic or pneumatic power for the plane's vital control systems. Although one remaining engine can supply these needs, it means that half the resources are lost on a damaged twinjet. As one US four-engine jumbo pilot who regularly flies the Atlantic commented, one engine failure left him with a large safety margin, but on a twinjet 'it spells a full-scale emergency'.

● Single-engine flight is not only slower but uses more fuel to cover a given distance, so that this and the unpredictable headwinds at lower levels call for even larger reserves of fuel to be carried on all flights, in case an engine should fail. The moving parts of a jet engine turn at vast speeds under immense pressures. A break-up can explode debris like a shrapnel burst around the wing and fuselage with imaginable results. Such 'uncontained' failures are closely monitored and present a continued anxiety to the safety regulators. Twinjets have fewer fuel tanks than jumbos, so that – if it happens – a rupture from flying debris in one tank must lose a larger proportion of the plane's remaining emergency fuel.

● Back-up emergency power is more limited. The auxiliary power unit – the mini-jet engine in the tail – is primarily designed to service the plane's needs on the ground before the main engines are started up. A secondary role is to provide emergency power in flight should the main sources fail. On a twinjet the APU becomes a vital piece of equipment, but after prolonged cruising at high levels it becomes 'cold-soaked' and cannot be started for some time after a descent, nor will most APUs start in the rarefied air at high levels. Single-engine flight may therefore have to rely on the remaining generator from the sound engine. If that source should fail – not an unlikely event – then only the bare emergency services can be kept going for a limited time by the batteries – a nightmare prospect for any airman in mid-Atlantic.

● When the crippled plane descends and heads for the nearest approved airfield (or 'alternate'), a number of imponderables face the crew. For example, alternates on the North Atlantic run will include Narssarssuaq in south Greenland, Sondre Stromfjord some 440 miles to the north, and Keflavik in Iceland (let's call them N, S and K respectively for short).

N boasts only primitive navigational beacons and no instrument landing system for bad weather on its single runway (strong winds can thus make it unusable). It can only be approached by a 28-mile flight up a narrow fjord beneath any cloud. Sheer cliffs ahead of the runway rule out turns or an overshoot that would permit a second attempt at the landing. (The pilot may, for instance, at the last minute find the runway obscured in a blizzard or simply be unable to make an accurate approach to an unfamiliar and tricky airfield.) At night or in low cloud the attempt would be even more perilous amongst the peaks that rise to 4,742 feet.

S is a fully equipped US military airfield but it lies about an extra hour's flying time to the north over high arctic mountains – longer in contrary winds. Again there is only one runway lying at the top of a fjord tightly flanked by mountains rising to over 2,000 feet.

Only K in Iceland provides the full facilities of an international airport, but all three are of course subject to the vagaries of sub-arctic weather. Pilots point out that the maximum crosswind component that a single-engine landing can safely withstand is about 17 knots, while crosswinds up to 35 knots are common enough. Besides the usual weather hazards of the region, winter brings snow- or ice-covered runways at the shortest notice. The odds are stacked against a safe emergency landing and a stricken plane low on fuel may not be able to circle and wait while the runways are cleared. A plane that reaches this high state of emergency is faced with what amounts to a crash landing.

Finally, besides the scarcity of weather data, conditions can change within the 60 or 90 minutes after an engine failure, so that a diverted plane with its emergency fuel spent may arrive at its last haven only to find that it has become weather-bound.

● It will be seen from the record of crises that follow that the mechanical reliability of the engines is not the only safety yardstick. Systems and human failures show themselves to be almost as great a threat to the safety of long-range twin operations. The engines may be blameless, but the planes still come down.

Fears allayed

Much indeed has happened to improve matters in the years since the IFALPA meeting. For example, it has since become a requirement for ER flights to carry APUs which are capable of either being started up at any height or being run continuously from take-off to landing. The package of ER extra equipment and modifications is becoming more standardized, as exemplified by Boeing's ER-model 767. The criteria for airlines seeking dispensations beyond the 60-minute limit include not only the level of the back-up systems carried, but also individual pilot operating experience, the routing and its known weather limitations.

A super outfit with fully-fitted ER twins may qualify for the 120-minute limitation or more – the FAA acceded to a claim by Pratt and Whitney and Boeing that the limit can be safely extended to 138 minutes. Now that a 3-hour single-engine flight time is available to US carriers who can meet the even higher safety standards this implies, virtually all the constraints on these twinjets can be lifted so that they may fly the longest routes along with the bigger planes.

In 1988, however, a study commissioned by the FAA itself raised doubts about the validity of its own calculations for the safety margins that it had adopted and has since recommended further research. This hitch came just as the FAA seemed prepared to extend the range limitations for twinjets by 50 per cent. The study faulted the FAA's criteria and publicly announced that 'a lot more statistical analysis and evidence is needed'. It went on to pinpoint the danger of unexpected headwinds which could 'increase the risks to unsafe levels' especially if these were compounded by secondary failures. Safety rules, it declared, should be based on distances that can be realistically flown, and not on time, as has been done up to now.

Yet in a fast-changing scene, ER twins may soon be free to fly on most world routes – unless, as Lord King feared, a mishap intervenes to shake public confidence and set the clock back. While the Americans are the pace-setters, the UK has cleared two ER-equipped airlines – Monarch and Air Europe – to operate under a 138-minute limit that is long enough to eliminate the last no-go area in the North Atlantic.

It is notable that so few British airlines have launched ER-twin operations, and even in the United States commentators report 'a market response that shows a growing ambivalence many airlines are displaying towards the concept'. Tadry Domagala, speaking for the engine-makers Pratt and Whitney, also sees a cooling off: 'Airlines don't know whether to commit fully or to buy 3- and 4-engined aircraft . . . it's a very volatile situation.' As the constraints take their effect in real operations, the number of airlines who think that they could operate EROPS twins profitably seems to be shrinking. Few of the nineteen airlines who are already doing so worldwide (1988) have attempted the North Atlantic, and many fly routes that can meet the old 60-minute rule and keep within range of safety.

In other words, for the newcomers the game may not be worth the candle. The law insists that the alternate airfields must be known to be usable for the necessary time-span before departures, and in winter there can be just too many days when Met forecasts will disqualify scheduled flights, at a costly penalty. The customer-factor is also beginning to tell. Recently an intending passenger telephoned a major transatlantic airline that is using a fleet of ER twins. He inquired innocently what type of plane he might be flying on, and the lady at the booking desk replied: 'Oh, don't worry, it's not two-engined.' When he asked the reason for this unsolicited fact, she replied: 'We've had a lot of people asking, you see – they don't seem to like the idea.'

As for the pilots, among fears that have been at least partly allayed is the likely elimination of Narssarssuaq as an approved alternate, thus positing the longer diversion to Sondre Stromfjord. APUs have been redesigned but their reliability has yet to be proved in practice: there is to be more stand-by generator capacity and both the FAA and the CAA require at least three separate electrical systems powered from independent sources. Since diversion times depend on the chances of surviving a fire on board, extinguishing systems must be beefed up, and safe flight must be ensured after the loss of any two hydraulic systems *and* either engine. Finally, the cooling systems for the avionics must be upgraded.

These changes take a lot of the sting out of the pilots' early fears and have begun to encourage the airlines towards bolder aspirations. In August 1986, some 62 Boeing ER 767s were already in operation with orders for a further 100 destined for 17 airlines. They included El Al, Air Canada, Qantas, Air New Zealand and American.*

Recent comments from the British Airline Pilots' Association show that pilots now believe that the UK requirements – which echo those in the USA – are 'adequate' but that there remains a vital need to test them out in practice before they are applied universally to passenger flights. However, they remain worried about the practicality of policing the new rules worldwide. Many of the mandatory safeguards are amorphous and variable according to weather and the proper functioning of a long list of systems and components. It seems that we shall have to wait and see what might happen if British twinjet crews of the major airlines were called upon to fly regular scheduled ER routes with passengers – quite apart from what their passengers might say about the prospect.

* As a guide in a fast-changing scene, in July 1989 there were more than 1,500 EROPS flights a week crossing the North Atlantic.

13 | 'Look – no engines!'

Since the idea of long-range twinjets was first mooted, there have been an astonishing number of engine-related emergencies on the new types. It is worth looking at the causes in some detail, especially those four that did lose *both* engines in flight. They will demonstrate why the soundness of the engines themselves – obviously a ruling factor – is far from being a complete measure of safety.

Causes of engine failure

Among the power-plants that have had most ER experience so far is the US Pratt and Whitney JT9D series on Boeing 767s, while Rolls-Royce offers its latest RB 211 variants and General Electric its CF6–80 series. While there is fierce competition between these and other jet engine suppliers over their comparative reliabilities, their profound technical arguments are less important to the user than the record of what has actually happened in practice.

The TWA B 767 that flew 400 miles on one engine over the Atlantic to land at Goose Bay in January 1988 (pp. 154–5) – the first event of its kind – had conducted itself in good order and observed all the new safeguards. Note that the FAA defines the EROPS portion of a flight as that segment during which the aircraft is more than 60 minutes' flying time, judged by its single-engine speed, from an adequate airport – 'a distance of about 400 miles'. So the Goose Bay excursion can be read in two ways: either as proof that such operations are now indeed safe, or as suggesting that – but for a bit of luck – it narrowly avoided ending in tragedy. Pessimists may think that the history of similar failures and the number of double-engine failures tends to support the second proposition.

Six months before Goose Bay, a Delta Airlines BE 767 with 205 people on board lost all power on both engines at 2,000 feet as it climbed out of Los Angeles over the Pacific Ocean. Now flying a

autopilot, from 41,000 feet over the Rocky Mountains. As it entered thunder clouds at 30,000 feet, its left engine surged and overheated: 18 seconds later the remaining engine followed suit. The crew had no choice but to face the 'horrifying thought' (as a colleague put it) of shutting down both engines to prevent damage or worse. As they did so, they broadcast a Mayday emergency call.

For one and a half minutes the airliner glided down in silence. There was no panic, and one passenger is said to have remarked to his companion: 'They tell us this is a mighty quiet plane, but this is ridiculous.' Meanwhile Denver radar controllers responded to the Mayday call by steering them down as best they could for the crash landing that now seemed inevitable. By this time the plane had dropped to 14,500 feet, having just passed over the Rockies that rise to 14,000 feet.

Shorn of all power from the engines, the crew struggled to start up the APU. They succeeded, and soon its back-up electrical power came on stream. This would allow them to re-start the engines without relying wholly on the batteries for ignition. While the general rule is never to re-start a faulty engine in flight for fear of fire or explosion, it is one that does not seem to address itself to the *force majeure* of engine-less flight. No surprise that the crew decided to risk starting up the engines one at a time. Luckily they succeeded, and without demur the engines took them back to Denver where they landed without further mishap.

Controversy surrounded a report made soon after the incident that the break in the main power supply from the two engines had wiped out the pilots' main flight instruments from the six VDU or television-type screens on which these are displayed on the B 767. If it had, it would have left the crew *in extremis*, having to rely on very restricted basic stand-by instruments to add to their troubles. First reports suggested that a significant time elapsed before the VDUs were restored, but others insisted that it had been no more than a momentary flicker and of no account. As the Safety Board's accident report makes no mention of VDU failures, the earlier reports must be discounted. Some 18 months later, the Board concluded that the malfunctions of both engines were probably 'the result of engine design and maintenance', and it accordingly issued corrective bulletins to both B 767 operators and manufacturers.

What had worried the Safety Board deeply was that less than a month before the Rockies adventure – on 27 July 1983 – another Boeing 767 had again lost both its engines at 39,000 feet. This time it flew a 100-mile powerless glide to crash-land on a disused Canadian military airfield at Gimli, Manitoba, narrowly missing the crowd attending a sports car meeting in progress on the tarmac at the time. The pilots of the Air Canada plane made a smart job of the powerless landing – a unique 'dead-stick' touchdown for a passenger jet – and none of those aboard was hurt, at least physically. As the shaken passengers described, it had been a ride of terror. 'Everyone was really scared,' said one. 'There was a lot of screaming and crying kids – it came down hard.' The nose-wheel tore off before the crippled plane braked to a halt only feet away from a group of tents where the race-drivers' families were camping.

The survivors owed a lot to the co-pilot of the flight. Marcel Quintal had flown from the field when it had been a military base and so knew its exact position and, equally appropriate for the occasion, he had ten years' experience as a glider pilot. This time the engines were not at fault. The plane had simply run out of fuel. It had been short-loaded for its flight from Montreal to Edmonton, because someone had confused imperial with metric measures (surely the classic case for those opposed to metrication – a bane to Anglo-US aviators the world over). The conversion from kilograms to pounds had left the tanks with only half the amount they should have carried for the trip. The engines had performed perfectly, but a powerless twinjet had bitten the dust just the same.

Nor were the engine designers to blame for a double-engine crisis on the Airbus A 310 flight which was reported anonymously by the pilots of IFALPA in June 1986. We are not to know the airline, but the engines in this case were General Electric's CF6-80 A3s. Although the A 310 is a contender for ER flight, the report is said to be 'officially of no significance' for the reasons that follow.

The flight on 26 June turned back 19 minutes after take-off when its crew saw warnings that both engines had run out of oil. As such indications had been found to be spurious in the past the crew, doubting their validity, decided that it was safe to overfly a nearby airport rather than land immediately and choose to return to base for a check. Soon both engines were showing zero oil pressure, but they managed to make home-base and land before shutting the engines down. Here it was found that mechanics had fitted the wrong size of

'O-ring' oil seals on the day before the flight. The seals on both engines were cut and pieces were missing, allowing the oil to be pumped out under pressure.

In order to prevent 'this kind of dual loss event', the airline now staggers its maintenance work on each engine so that such fitting errors are confined to one engine and not both. But the pilot-reporter goes on to comment that since the dual-failure 'event' did not result in an *in-flight* shutdown, it is disregarded by official statistics. Had the oil leaks happened rather further from home, perhaps over the cruel seas, one can presume that the consequences would then qualify as an official event with a vengeance.

The British Midland tragedy

Similar but more vivid speculations about the safety of twinjets followed the fatal crash of a British Midland Airways Boeing 737–400 on the night of Sunday 8 January 1989. One of its two engines failed soon after Flight BD092 had taken off from London's Heathrow airport en route for Belfast, Northern Ireland. The crew began an emergency diversion to East Midlands airport near Derby, but it crashed short of the runway onto the M1 motorway, killing 44 people and injuring 82 others. Both its pilots survived.

Ironically, the disaster came just as the US Federal Aviation Administration was about to allow American Airlines to fly its ER twinjets further from airports than ever before by extending this to 180 minutes flying time. But did the British Midlands crash have much significance in the argument about ER twinjet safety and reliability?

As a short-haul twin, the B 737 is neither in the league of the ER types nor did it pack the added back-up systems required on ocean routes. Most engine failures occur at take-off or in the climb – as had happened on Flight BD092 – and are much less likely to do so at the lower power settings used during a steady cruise over ocean and remote routes. So the only relevant factor to be gleaned from the accident in a discussion about ER safety seems to be that an engine failure on a twin had led to a fatal accident. Yet the events of that 32-minute journey to disaster uncover human factors that can come into play on any flight deck.

According to airline regulations for such a single-engine

171

emergency, the pilots of the stricken British Midland were attempting a standard night-time descent down the instrument landing system (ILS – see pp. 217–18) toward the runway. Small adjustments of engine power are needed to keep the plane on the glide path and, at the last minute, when the pilots called for power from the apparently sound engine, it failed to respond. For an instant, it must have appeared to them as if both engines had failed at the crucial moment – and indeed, the consequences were no different from a total power failure. We shall see whether or not this was so. Apparently powerless, the plane sank below its safe approach path and crashed onto the motorway only 1,000 yards from the runway threshold.

One of their engines – the right – had been perfectly sound and could have saved them, but they did not know it. This fatal misconception had built up in their minds during the 17 minutes since the engine fault occurred near the top of their climb to 30,000 feet out of Heathrow. For a variety of reasons which are still being examined, both of the pilots had identified a failure in the right-hand engine, when in fact it lay in the left engine. Consequently, they had then shut down the right engine and relied on the left (and damaged) engine to fly the descent.

As normal, they had reduced power for the long descent from 30,000 feet. The effect of this was to mask the vibration and danger signals from the flawed engine enough to allow them to persist in their wrong diagnosis, continuing to believe that the left engine was the sound one. When they were 3 miles from the runway and at a height of 900 feet, they called for power from the left engine, which of course failed to deliver. Now about 90 seconds before touchdown, the crew may either have thought they had a double engine failure, or realized their mistake. *In extremis*, the captain called for the right engine to be re-started, but it was too late to check the descent and clear the boundary bank of the motorway that lay ahead.

The events that led up to this mistake in identifying the failed engine need not concern us here. Keeping speculation to the minimum, it is enough to say that many of the accidents already cited show how easy it is for one person's mistaken train of thought to infect two or more minds working under stress. Such a group-fixation on one apparent symptom tends to reject other and contrary signs: it fits the medical condition known as a psychological 'set'.

There is, of course, a mass of instruments and warning devices to assess when an engine fails on a modern jet airliner and, like all mechanical things, they can be fallible too. But as all aviators know from their early training on smaller aircraft, it is suprisingly easy to identify the wrong engine when one of them suddenly fails, and a lot of practice goes into overcoming this hazard.

The instructor would hold a card over the throttle levers to mask them from the pupil and then surreptitiously cut one engine during a take-off or landing. But so many fatal accidents were caused by pupils shutting down the remaining good engine, that the drill is now only practised by simulating it at a safe height. (The test can be eminently more exhilarating at night when the trainee is flying 'blind' and is only permitted to see a carefully limited number of basic flight instruments on which to control the results of an engine failure.)

A fool-proof method taught to trainee pilots on light-twin aircraft – but barely applicable to big jets – worked well. When an engine suddenly stops, let us say the left one, the plane naturally slews left towards the dead engine. The natural and necessary reaction to this is to apply strong foot pressure to the right rudder pedal to stop the plane from turning and rolling to the left. The left foot therefore feels slack on the other rudder pedal. The trick taught was to lift up the disengaged left leg, slap the knee and shout, 'Left leg, Left engine!' But as can be imagined, jet airline pilots have rather more to contend with.

The final relevance of the British Midlands tragedy is simply that, for whatever reasons, two pilots found themselves in a situation in which both engines had apparently failed, and another powerless twinjet crashed into the ground. If they had had a third or fourth engine, it could have been prevented. However less likely it is that this might have happened on an ER flight, and remembering all the curious concatenations of human behaviour, it would be tempting fate to expect every single-engine emergency flight to reach its haven in perfect order and without incident.

Flight BD092 also puts another question mark over the tendency to accept engine reliability as the ruling criterion of safe ER operations: whether engines break down or are shut down by the crew – rightly or wrongly – the consequences are the same for the passengers aboard. In reality, the risks of flying with two engines must be judged on the likelihood of a total loss of engine power, however this may come about.

An ominous sequel to the British Midland crash came five months later in June 1989. In one weekend, two Boeing 737-400s suffered single engine failures identical to that which triggered the disaster in the January before. In each case, blades from the front fan of the French-manufactured CFM56-C3 engines had broken up and been sucked into the high-speed turbines. The fault recurred on a British Midland Airways sister-ship of the crashed plane, on the same route from Heathrow to Belfast, and again on a Dan-Air charter flight from London Gatwick to Minorca in the Mediterranean.

Both planes landed safely, but all B 737-400s fitted with the same CFM engines were grounded until safety modifications had been devised and carried out.

The record so far underscores the difference between the reliability of the engines judged purely on proven design and operating statistics, and routine (and mandatory) *shutdowns** by crew when an engine or systems fault declares itself or crises are caused by human error. Plane-makers tend to recite the former as proof of their wares, while pilots and safety watchdogs insist that the interaction of all these – irrespective of the cause – is the true measure for the safety of long-distance flights on two engines. For this reason I contend that, while sheer engine reliability is obviously the prime target, the plane-makers' statistics – however valid in their restricted context – can not only be misleading by themselves, but do not deserve anything like the attention they have won for themselves. We as customers need to know the safety of the whole package under the test of experience, however bitter it may be.

Multi-engine shutdowns

Whatever confidence may be placed in the twinjets, it is not strengthened by a number of total engine failures that have occurred in both three- and four-engine planes. Even Captain Moody's four-engine failure over Jakarta (pp. 63–8) is not unique: in 1974 all four engines of a British Super VC 10 stopped while it was cruising high

---

* 'Shutdown' denotes the deliberate closing down of an engine by the crew following a real or spurious warning of an abnormality in a number of engine or engine-related systems, and may be a mandatory response in line with standing operating orders. This is distinguished from uncontrolled loss of power caused by an actual mechanical breakdown of an engine.

above Japan because the co-pilot had fed them all from one tank until it ran dry. This reliable old plane is fitted with a ram air turbine (or RAT) that can drive a small emergency generator. Normally stored inboard, it can be lowered to windmill in the slipstream from the belly of the plane. During a 9,000-foot plunge, the VC 10's crew managed to deploy their RAT and so gain enough power to energize the fuel pumps and restore the feed to the engines, all of which then responded.

A similar spot of finger-trouble on the fuel switches shut down all the four engines of a Boeing 747 jumbo during a flight from Boston to London in July 1979, at a point 80 miles out to sea between Newfoundland and the toe of Greenland: it fell 8,000 feet from its 35,000-foot cruising level before the mistake was recognized and the engines then coaxed back into life. Other fuel troubles have had less happy endings, such as the United Airlines four-jet DC-8 that simply ran out of fuel and crashed short of Portland airport, Oregon, in December 1978. Remarkably, only ten people died although many others were seriously injured.

The tri-jets have also had their share of total failures. Everyone survived the event when all three engines cut out successively on an Eastern Airlines Lockheed L-1011 during a short hop to Miami from Nassau, Bahamas, in May 1983. The big jet descended from 13,000 to 4,000 feet above the sea before one engine was re-started and – by the narrowest of margins – managed to squeak into Miami airport. The cause of that event foreshadowed the oil leaks that struck the Airbus three years later (pp. 170–1). This time the mechanics had simply omitted the vital 'O-ring' seals altogether, so that the oil streamed out from all of its three Rolls-Royce RB 211 engines within minutes after leaving Nassau.

The significance of all these multi-engine shutdowns or stoppages is more than it may appear, because the state of an emergency on a twinjet is predicated on the loss of only one engine, not two. The case of double-engine failures does not feature in the prospectus, for obvious reasons. Yet single-engine failures on twinjets, if not ten-a-penny, are legion. Leaving aside the list of double-engine failures, the eight instances of single failures that have featured in published accident reports for the five years 1982–7 alone are regarded as a conservative total. Taking these alone, they group themselves as follows: Airbus (4), B 767 (1), B 757 (1), MD-82 (1), DC-9 (1). One

175

other British B 757 of Monarch Airlines suffered, not an engine failure, but a four-minute double electrical power failure that temporarily robbed the crew of their flight instrumentation. By any standards, it is not a good record for the reputation of aeroplanes limited to two engines.

Regardless of type, double-engine failures continue to happen too often for comfort. In May 1988 a three-engine failure struck a United Airlines Boeing 747 bound for Tokyo. The first failed 75 minutes before landing, the second 30 minutes later and the third just as it began to make an emergency landing at Narita airport, 40 miles short and east of Tokyo, where it scraped in with 239 passengers on board. Faulty fuel supply was suspected. Others that have come to light include a Swearingen Metro commuter jet that crashed after both engines flamed out during an 'inadvertent' penetration into a thunderstorm, killing 13 people. The NTSB's recent comments on this accident in 1985 blame 'massive water ingestion' into both engines.

Confronting the elements

Unlike airline travellers, there are some intrepid spirits who positively relish flying over the freezing seas and icy fjords – even on one engine. The story of such an attempt in January 1986 provides a vivid insight into the contest between man and the sub-arctic elements – a contest that two young Frenchmen were to lose.

The pilot had flown the route before and held a commercial licence. He had just bought the six-seater Cessna Centurion in the United States and planned to fly it back to France, where he owned a flying outfit. For its size, it is quite a sophisticated machine, with a pressurized cabin and a single turbo-piston engine, but its limited fuel endurance called for short hops – Newfoundland, Greenland, Iceland, Scotland. His co-pilot was a well qualified flying instructor but without a commercial ticket. As they carried no long-range (HF) radio, they were restricted to flying the route under visual flight rules (that is, no blind flying in cloud or at night).

They took off on a January morning from Goose Bay, Labrador, on their first 700-mile leg to Narssarssuaq near the southern tip of Greenland, where the forecast was windy with clear visibility – but it added a warning of temporary low cloud and snowstorms developing by their estimated time of arrival (ETA) which lay five hours ahead. They filed their fuel endurance time as seven hours.

After a couple of hours' flying they heard by radio that the weather at Narssarssuaq was indeed deteriorating – a 44 m.p.h. partial crosswind with rain, snow and ice-pellets. A few minutes later a warning came through of severe icing conditions on the route ahead. Now the weather at Narssarssuaq was below the legal safety limits for a landing. The chart of the landing area now spread out on their knees carried a warning that approaches should only be made if the cloud base is not below 4,000 feet and the visibility at least five miles – and then only if pilots had 'a good knowledge of the local topography and meteorological conditions'. The chart also told them, as we already know, that the mountains surrounding Narssarssuaq and the narrow fjord by which it must be approached rose to 4,742 feet. Now the forecast for the airport predicted 'sky obscured' – that is, fog, or cloud down to ground level.

A landing was clearly ruled out, as well the chances of landing at a nearby alternative landing field to the north which was equally weatherbound. The pilot radioed air traffic control to say that he would therefore divert to Reykjavik airport in Iceland, giving his ETA there as 17.17 – which would be $5\frac{1}{2}$ hours from take-off at Goose. He added that his fuel would last until 18.28, thus giving him a good hour's safety margin.

His short-range radio was soon beyond the usable range to the ground controllers, but at 16.58 he managed to pass a message to them via the crew of a British Caledonian plane flying overhead and who relayed his call: he told them that he was 150 miles west of Iceland and now revised his ETA for Reykjavik to 18.28 – but as he did so, towering cumulo-nimbus storm clouds were building up over the Icelandic haven. At about this time he managed to report that he was low on fuel, now with only an hour's endurance left. Next, at 17.12, Iceland radar controllers spotted him on their screens 215 miles east of the airport.

Now desperate, the two Frenchmen tried to avoid unexpectedly strong headwinds at their 5,500 feet cruising level by climbing to 15,000 feet but, unknown to them at the time, the upper headwinds were no less severe. Although the pilots had not declared an emergency at this stage, the Icelandic authorities took the initiative and triggered the search-and-rescue alarm at 17.23. A USAF C 130

(Hercules) and surrounding ships were immediately diverted by the controllers to a point ahead of the radar blip they knew to be the Cessna. By now the weather in the search area had blown up to a 45 m.p.h. wind with 23-foot waves on the ocean swell.

At 17.51 the Cessna began a descent from 15,000 feet. Twenty minutes later its fuel ran out – 6 hours and 24 minutes after leaving the safety of Goose Bay. By this time the C 130 had made radio contact with the Frenchmen. In the deepening dusk, the rescue plane dropped flares as the powerless Cessna prepared to ditch, assisted by a talk-down from their US rescuers who were now circling overhead. Just before the final splash, the C 130 commander told them: 'You're looking good . . .' He saw that the Frenchmen had made a successful ditching between the huge waves, but when a USAF helicopter joined the search three minutes later there were no signs of the two occupants. After five minutes the plane was lost to sight. Despite the gathering darkness and foul weather, the search for survivors continued for six hours but it proved to be fruitless.

The French pilots, who had they been granted another seven minutes of flying time would have reached safety, were never found. The report of the accident confirmed that their position estimates had been too optimistic by 100 miles and more, mainly due to the erosion of their safety margins by the unknown winds and the vicissitudes of the sub-arctic climate.

The validity of any analogy between this flight and the safety of twinjets operating over similar routes centres on two factors. The first is the number of days in the year on which it is known before take-off that the emergency airfields are not weatherbound and, more importantly, that they will not become so during the vulnerable sectors of the whole flight-time. In winter, it has been roughly calculated that on average such days may be counted on the fingers of both hands – a statistic that may explain the recent loss of enthusiasm shown by scheduled airlines.

Ditching

The second factor is the conditions facing passengers in the event of a real emergency – since a ditching cannot be ruled out however remote it may seem to be on present form. The chances of a survivable ditching are only too clear unless or until new survival techniques can be developed. Not only would the French pilots have been required to wear life-jackets (whether they did so is not on record, but it would be unimaginably foolhardy to disobey the rule), but ejecting from the front seats of a light aircraft is a lot faster than queueing in the aisles of a ditched airliner. As required, the Cessna also carried an automatic emergency locator transmitter that duly signalled its position to the rescuing craft for 15 minutes after the crash. But even this was to no avail in the darkness and freezing ocean swell. There are just too few minutes to beat the sudden and lethal effects of immersion in cold and tempestuous seas.

Ultimately we have to face the unspeakable possibility of a ditching in mid-Atlantic or in other frigid waters – a prospect that must make the flesh creep, if not shiver. Most survival experts believe that it would be catastrophic, unless luck allowed it to happen very close to inhabited land or near a ship: even then the odds are poor. One only need imagine 20-foot waves and a 30-knot wind, perhaps in sleet or in the dark or both. . . . Whatever survival kit is available on board, the chances of deploying it and loading several hundred passengers safely into rafts in such conditions are thought to be minimal. In those circumstances human survival time in the sea – a matter of minutes – becomes academic. One sardonic search-and-rescue expert even regards the provision of sea-survival equipment as not much more than a sop to the passengers. (He may or may not have been the author of the black joke, voiced in the magazine *Flight International*, about the aircrew flying a twin-engined plane above the ocean, in which the captain asks what 'EROPS' stands for. His co-pilot obliges with 'Engine Running or Passengers Swimming'.)

Happily there have been no full-scale and remote ditching accidents in the jet age and so there are no practical lessons to learn from. When a two-engined DC-9 ran out of fuel and ditched over the warm waters of the Caribbean in May 1970, a total shambles ensued in the cabin. Life-rafts were inflated inside and jammed the doors and only one escape chute was deployed. Twenty-three of the 70 on board died as the plane sank 5,000 feet to the ocean bed, only 30 miles from the Virgin Islands. All 102 people on board perished offshore from San Juan, Puerto Rico, in the same year when another DC-9 lost power after take-off and crashed two miles from the shore. In 1978 three people died in a Boeing 727 that crashed 3 miles short on the approach to Pensacola, Florida – in only 12 feet of water.

Striking water at speed is like hitting concrete. The growing number of ER twinjets now flying across the oceans has naturally revived concern about the behaviour of an airliner when it hits the sea. Certifying authorities require the ditching abilities of each type to be 'proven' either by simulated tests using scaled-down models, or by mathematical predictions based on analysis and extrapolation. The expense of ditching trials with a real aeroplane, as Airbus point out, would be prohibitive. Yet it is a measure of the new concern about twinjet safety to see the search-and-rescue experts, who are alert to the prospects ahead, thumbing through the records of ditchings from the pre-jet era with a revived interest.

It remains for the industry and the regulatory authorities – and not the confederacy of aviation interests alone – to prove to its customers beyond a peradventure that twinjet long-distance flying is at least as safe as it has been in the bigger planes of the past. This may well be achieved in the future, but for the moment we, the passengers, and the pilots who fly us are justified in reserving our judgment. To paraphrase Lord King, a 'not proven' verdict is not enough: it needs to be proved not on a balance of probabilities, but beyond a shadow of a doubt.

14 | Ball of fire

Seasoned business travellers who have come to use airliners as intercontinental taxis are perhaps the people least likely to be perturbed by the thought of accidents. The truth that familiarity breeds contempt may be another way of saying that the human brain takes care that you can't go on being frightened all the time. But it is safe to say that few people apart from drunks and fools wholly escape that occasional twinge of apprehension that intrudes into our reflective moments. It may be no more than a primeval reminder that flying is a most improbable adventure for our wingless species.

To be strapped helplessly into a crowded and flimsy aluminium tube containing tons of kerosene fuel, and then to be zipped at 500 m.p.h. through skies occupied by other similar projectiles can suddenly grip the mind as sheer insanity until the next drink is served to restore some twentieth-century sang-froid.

My particular version of that momentary nightmare is a fire on board. Attempts to dismiss this as a mere phobia are not helped by the facts.

Total disasters in which everybody is killed – those dubbed as fatal non-survivable crashes – are of course much fewer than fatal but survivable accidents, in which some die and some survive. In these more common events it follows that some of those who perish could or should have escaped alive, but did not. The human psyche shows wisdom when it prompts a fatalistic response to the rarer prospect of a non-survivable crash, whether it be crashing into a mountainside or sudden in-flight oblivion by a terrorist bomb. Your fate is then truly in the lap either of the gods or Satan himself.

The more worrying and more likely scene is having to play a part in a 'survivable' accident. It forces questions about how you may react when you know that your own behaviour may well determine whether you live or die. Nervous thoughts flit through the mind about emergency exits, surrounding panic, the effects of the impact, fire and smoke.

Experts assure us that 80 per cent of all passenger aircraft accidents occur within 1,000 metres of the airport of departure or landing. So it is some comfort to know that fire and rescue teams are usually close at hand. A British company in the fire-fighting business claims that 99 per cent of passenger survivals in such accidents occur when the aircraft is within half a mile of an airfield; and that on average 95 per cent of the passengers survive if there is no fire. Much more significantly, when there is a fire only 35 per cent of them survive. If fire is such a killer, how often does it happen? The answer is a sobering one.

For some time the trends have shown that more than half such crashes result in a fuel fire, and in most cases it spreads to the passenger cabin within seconds. In broad terms it has remained true that in 'fatal but survivable crashes' the main cause of death is not physical injury from the impact of the crash, but fire – or more exactly, the consequences of a cabin fire. An early but detailed study by the International Civil Aviation Organization (ICAO) in 1980 showed that in a review of 272 fatal accidents, 782 people out of a total of 1,018 deaths were killed by fire – over 70 per cent. More recent surveys which are mentioned later reinforce that high risk.

Why are casualties so high in fires?

Polyurethane foam and other plastic cabin fittings are the killers. A few moments after a fuel fire has started (typically between 10 and 30 seconds), the burning foam-filled seats and plastics emit lethal hydrogen cyanide and carbon monoxide gases. The oxygen content of the cabin air falls as the carbon dioxide level rises. Belching black smoke and solid particles obscure everything and choke the mouth, nose and eyes. Safety rules demand that in emergencies airliners must be able to evacuate the passengers within 90 seconds, but time and time again this has been proved to be not long enough amid the confusing blackout and shock of a blazing inferno.

Traditionally, the safety tests on foams and plastics used in aircraft have been based on small laboratory work-bench samples of the stuff. But other trials completed before 1983 by the UK Fire Research Station of HM Factory Inspectorate had already shown that such bench tests were inadequate. In aircraft, the dangers are enormously enhanced because the flammability of foam cushions becomes near-explosive when they are massed in an enclosed tunnel of metal and exposed to the high temperatures of a fuel fire. The resulting cabin heat was found to reach 1,000 degrees Celsius, and surprisingly even higher when so-called 'self-extinguishing' foam fillings were tested – enough to melt a steel-plated floor. Within two minutes toxic gases and oxygen depletion reached peak values and made the cabin unsurvivable. The pain threshold comes at 45 degrees C and at 70 degrees C the survival time is one second. The report of the trials warned that the impact of the gases on passengers 'represents an important additional hazard which may not previously have been recognised'. And the hazard has continued to be with us for far too many years after it had been recognized.

The long-known dangers of modern synthetics in the home are a sorry enough tale. But the consequences of a fire on land or even at sea are nothing compared to a fire in the air, and it is preposterous that foam seats and plastics have been tolerated in airliners ever since the materials became available nearly a couple of generations ago. Consider the safeguards set for cinemas, theatres and buildings used for public assembly. John Lodge, a senior fire consultant, points to 80 inches of print setting out the mandatory requirements for such public places which, he says, even government officials admit 'are more demanding than those which apply to passenger aircraft'. He rightly describes this as a blinding glimpse of the obvious, 'since to impose normal building specifications on airliners would ensure that they never fly again'. Not many travellers are aware that an airliner, as a place of public assembly while it sits silently at the loading bay, is that much more hazardous than their local cinema or theatre. All the more reason, then, to do the most that can be done. Lodge relishes an imaginary scenario whereby an intending builder of a cinema is seeking a licence from the public authorities. The entrepreneur explains that it will be constructed of aluminium sheeting in narrow tubular form, bolstered round with hundred-gallon tanks of aviation fuel connected to three or four mighty jet engines capable of running at 1,000 degrees Celsius. The seating? Polyurethane foam, of course.

Ever since the day of Orville Wright air passengers have accepted that flying has its inherent dangers, but packing them up in highly flammable foam should never have become one of them. In the years since we at the *Sunday Times* conducted our primitive fire-tests on safer products, many more substitutes – some for the foam itself and others for the cabin's plastic panels and surface fittings – have come onto the market with claims to halt the onset of a cabin fire and lengthen the escape time available.* Although the safety roles of such materials differ, most of them are neither prohibitively costly or heavy (one even claims to reduce the weight of each seat by 0.8 kg), so that for some years now the industry's claim that no suitable substitutes were commercially available has been a false one.

Delay, complacency and prevarication

This was the case at the time of the 1980 Riyadh fire (p. 1–11), and the fact was reaffirmed in the wake of the next major fire disaster to hit the headlines in September 1982. The loss of 55 lives in that fire is particularly poignant because it was triggered by a relatively minor hazard – the bursting of a nose-wheel tyre – that need not have had the results it did. It happened to a Spantax DC-10 just as it was about to lift off from the runway at Malaga airport in Spain. Later analysis showed that at a speed of 160 m.p.h. the burst nose-wheel tyre was already practically off the ground and its pilot could have continued to take-off safely, to be followed by a less risky and controlled landing. But he decided to abort too late and overran the runway at speed, crashing through a perimeter fence before the plane broke up in a ball of fire. Once again most of the victims died in blinding black smoke from toxic fumes inside the cabin before they could fight their way to the exits. The bodies of those who were last to succumb were found piled on top of one another against the closed doors.

The report of the accident shows that much of the investigators' attention focused on the fallibility of aircraft tyres, the rights and wrongs of aborting a take-off so late, and the airport's culpability for allowing a solid concrete structure close to the perimeter which in the event had caused the final break-up. Precious little was said or done about the proximate cause of death, nor was there any backward look at Riyadh or the catalogue of earlier crash fires to identify the common danger. Officials were then content, when challenged about

* Some of those more recently announced include, for example, Metzoprotect (Europe), Beaverco, Polyvoltac and Solimide (UK) and Ryton PPS (USA).

the part played by dangerous cabin material at Malaga, to repeat the formula that 'the matter is currently under review'. Just how long had it been under review?

One of the starkest warnings came as long as 15 years ago when a cabin fire broke out on a Brazilian Varig Boeing 707 minutes before it came in to land at Paris Orly airport in 1973. Within sight of his destination, its captain executed a perfect landing in a ploughed-up orchard (surely a world record for any big jet). The feat was the more remarkable because the plane was hardly damaged and the fuel tanks, for once, remained intact so that there was no external fire. But when the flight crew clambered out of the cockpit and opened the front main cabin door, they were met by billowing smoke and silence. Precious little life was left inside.

A blaze in the rear lavatory had spread to the seats in the last few minutes before touchdown and turned the cabin into a gas-chamber in which 123 passengers out of 134 had been rapidly asphyxiated.

This, as well as other portents that followed it, went unheeded until the same menace struck again seven years after the Varig fire, and with a vengeance. Every one of the 301 souls inside a Saudi Lockheed 1-11 were gassed to death as it burnt on the runway at Riyadh in Saudi Arabia in August 1980. The jumbo had left the airport a few minutes earlier, and from a distance of some 50 miles away its captain radioed that he had a cabin fire on board and was turning back to the airport. Eight minutes later he made a seemingly unhurried landing and drew the plane to a halt. Enigmatically, he calmly radioed the control tower: 'We congratulate the passengers for a safe landing. We are now trying to open the doors (from inside) to let out the passengers.'

They were his last words. By the time rescuers opened the doors from the outside, all the 301 passengers and crew were dead. In the words of a medical official, they had all been killed by 'a lethal cocktail of poisonous fumes' from the burning seats. Airlines flying other L 1-11s were shocked by the totality of the event, and a worldwide scrutiny of the plane's systems ensued in a search for the cause of the fire. Any mention of the cause of death was left to medical and legal functionaries.

Such classic fire tragedies come from a formidable list of others in which many passengers have suffered a like fate. They include the 23 who died after an in-flight fire swept through the cabin of an Air

Canada DC-9 cruising near Cincinnati in June 1983. The pilot made an emergency descent and landing but they died before the blazing plane could be evacuated. Other accidents listed under headings other than fire can hide the fact that it was indeed the actual killer.

The prodigious head-on collision between two Boeing 747 jumbos on the fog-bound runway at Tenerife's Los Rodeos airport in the Canary Islands remains as the world's worst aircrash since it happened in March 1977 (and long may it remain so). The combined deaths of 583 people in the KLM and Pan Am planes might have been expected to have been caused by the sheer impact between 700 tons of hardwear meeting at speed. But witnesses from the 61 survivors from the Pan Am plane (there were no KLM survivors) told that the impact had not been at all severe. Both airliners burst into balls of fire on the dark and mist-swept runway. Rescue teams groping towards them first saw a huge shape 'totally enveloped in flames, the only visible part being the KLM's rudder'. Describing the fire on the KLM plane, the accident report later told how 'an immediate and raging fire must have prevented adequate emergency operations because all the aircraft's evacuation doors remained shut even though the fuselage was not deformed significantly'.

When the inferno was finally extinguished, the state of the bodies prevented autopsies on some of them but it was inferred that most if not all had lost consciousness from inhaling toxic fumes before anyone could open the doors: the grisly effects of fire came later. The respective death tolls were all 248 KLM passengers and crew and 335 on the Pan Am plane. Survivors from the latter told how, classically, flames under the wings led to an explosion immediately followed by a fire inside the cabin.

A bitter international controversy followed about the factors that contributed to the disaster, but the consensus finally ascribed it to misunderstandings by radio between the airport traffic controller and the KLM pilot, who believed that he had been given clearance to take off while, unknown to him, the Pan Am plane was taxiing up the runway towards him in the fog. Whether or not the crash was survivable for the occupants of the KLM plane will never be known, but the results of the cabin fire in the Pan Am jumbo retell the familiar story. Any proposals for action drawn from the main cause of death are singularly absent from the section headed 'Survival' in the official report of the accident.

In the twenty years from 1966 to 1985 there were 74 airline fire accidents in the western world in which 2,497 people died from fire – about 29 per cent of the total number of occupants.* This sequence of tragedies offers an eternal message, not, one hopes, any longer directed solely to aircraft fires of the future, but because it reads across into many other chapters of air safety as an emblem of delay, complacency and prevarication. Nobody can say for certain how many of the 2,497 people killed in the last two decades of airline flying were choked to death by burning toxic cabin materials, but the post-mortems in known cases suggest that it is an overwhelmingly large proportion of the total. Fire experts warned in vain until, after nearly twenty years, the public got to know and charged the politicians. Only then did they win protection. For that reason alone it seems a tale worth the retelling.

Despite this series of catastrophic crash fires, airlines jibbed at the expense of fitting the modest safety expedient of more fire-proof seat covers until the authorities, reacting to the public outcry after the 55 deaths at Manchester, finally forced them to do so in 1987. According to the UK Civil Aviation Authority, 500 of the 2,497 people killed in the last two decades of airline flying would have been saved if the new seat covers had been fitted. Safety officials figured that the cost was between £100 and £200 a seat, or 10 pence per passenger per flight. They estimated that the bill for British Airways, for instance, would be around £12 million. One senior official figure wryly commented that that airline 'had just spent £25 million on jazzing up its Club Class seating'. But he admitted that it was not entirely a fair comment because British Airways – unlike other carriers – had pioneered the use of the safer seat covers months before the law had required it to do so.†

Once again it was to be shown that bureaucracy reacts with the most alacrity when it feels the shock of a fatal event on its own doorstep. Its belated response does at least suggest that the sacrifice of lives at Manchester airport was not wholly in vain. Although many worse fire disasters had happened before, those families of August

* These figures come from the UK Civil Aviation Authority.

† Tougher 'fire-blocking' seat covers that encase foam will delay but not prevent combustion. This is certainly a step forward but just how long it will extend evacuations in a real cabin fire remains a matter of controversy, and many fire experts remain sceptical about the claims made for them.

holidaymakers whose trip suddenly ended in a searing inferno before their plane had even got off the ground stunned readers of the national headlines for many days on end as the inquest into their deaths began to uncover the causes.

The Manchester fire

A mood of elation and anticipation turned to death and terror within a matter of seconds after the British Airtours Boeing 737 began to roll forwards at 7.13 a.m. on Friday 22 August 1985. The Corfu-bound plane with 137 people on board had accelerated to over 100 m.p.h. down the runway when, just before reaching take-off speed, its port engine exploded and burst into flames. The pilots, unable to see the engines from the cockpit, only heard a loud thud and at first thought it was a bursting tyre or possibly a bird striking the fuselage. The aircrew, following their emergency drills impeccably, aborted the take-off and managed to slow down in time to turn off onto a taxi-way in order to free the runway for other traffic. Aborting a plane at the last minute calls for split-second skills, but on that morning the crew appeared to have coped perfectly as they drew to an orderly halt. 'Don't hammer the brakes,' Captain Peter Terrington had warned his co-pilot, thinking that a flat tyre might overheat. Then the pace begins to quicken.

Five seconds later the fire alarm bell rings in the cockpit but the crew still cannot see its source or extent. Air traffic men in the control tower see the flames and exchange radio calls with the crew.

Captain: 'It looks as though we have fire in number 1 [port engine].'
Tower: 'Right. There's a lot of fires. They're on their way [the fire teams].'
Captain: 'Thank you. Do we need to get passengers off?'
Tower: 'I would, starboard side.'
Captain announces to passenger cabin: 'Starboard side, passengers off, please.'

These laconic but polite exchanges show that it was impossible for the flight crew to realize the full scale of the emergency that was unfolding behind them. Their cabin crew, who could see it, are trained not to interrupt the pilots while they are dealing with an emergency.

188

By this time the passenger cabin was engulfed in flames with temperatures rising to 500° Celsius. Fire and rescue teams who reached the scene within 22 seconds said they had never seen anything like it. Why had it all happened so quickly?

First inquiries revealed that the engine explosion had severed its fuel supply pipe, which continued to pump fuel at a rate of 250 gallons a minute into the engine fire behind the wing, creating a searing torch that burnt all in its path like a vast pressurized blow-lamp. Metal on the port side of the cabin simply dissolved as the foam-filled seats and plastic cabin fittings billowed black smoke and toxic fumes.

One passenger who narrowly escaped from the holocaust that followed lived to tell a tale that many less fortunate victims must have suffered to extinction in the long record of past airliner fires. Soon after the event Royston Metcalf, a dental technician from Derbyshire, told how in the first few seconds he felt his flesh creeping under the intolerable heat as the dense smoke billowed down from the ceiling of the cabin to where he sat in row 9. 'Within four seconds,' he said, 'it was as black as night. I turned to get my fiancée out. She said: "What about my handbag?" I told her to forget the blessed thing.'

Tongues of flame were now shooting down the fuselage as the smoke and acid soot choked his ears, eyes and nose. People around him were on fire but, unable to see in the blackout, they were moving the wrong way in search of the exits. 'I saw one lady who had just had her hair done, it must have been lacquered, because all of a sudden – snap, the lot went up. In panic she ran the wrong way.' With the aisle now jammed with bodies, others stampeded over the backs of the seats. Unable to see anything, he found himself wedged in the aisle, but with some presence of mind he began to count the rows as he felt his way forward to row 4. His hand met someone's head and knocked off their spectacles – 'I thought she was a goner. Then my left foot got caught under somebody's armpit or crutch and I fell into row 2.'

By now the smoke was pushing down nearer to the floor and he began to choke as flames roared overhead. Recalling the moment, he said that he had no sense of smell, taste or touch but just a drifting sensation 'for what seemed like about a week'. Then he heard a voice calling out his name, telling him to get low onto the floor where there was still some air: 'I realized it was my own voice and that I was actually on the floor. I always kept a hanky in my shirt pocket and I picked it out to try and breathe through it but my mouth was full of this filth from the seats – I put my finger in my mouth and brought it out like crumbled Oxo cubes – I couldn't breathe, I'd lost my glasses – I pulled the hanky over my right eyeball to get rid of this muck. By doing so I saw a little light as big as a postage stamp towards the bottom left and I made a bee-line for it . . .'

The greatest personal tragedy for Royston Metcalf was that his fiancée was not among the survivors.

Outside the ordeal in the cabin, fire and rescue teams pumped fire-quenching foam on the remains of the port side of the fuselage. Amazingly enough, many seats away from the centre of the blaze had remained untouched. When the flames were under control, the sight of the smoking carcass of the airliner told that the rescuers – including the crew – had performed a feat in saving as many as 82 lives. Post-mortems confirmed that the 55 who stayed inside had died from the inhalation of fumes before the effects of fire reached them. Unlike Royston Metcalf, they had lost consciousness sooner, and with it, the ability to escape.

Two of the crew who perished – Sharon Ford and Jacqueline Urbanski, aged 23 and 27 – were stewardesses who were operating the doors and could have been the first to escape. Unforgettably, they chose to return to the cabin to help save their passengers.

Another survivor, a student from West Yorkshire, described the panic of the scene inside. She had noticed a lady standing and holding a little girl by the hand. The lady was telling the girl: 'Push, Becky, push.' The student explained how the crush of the people behind her caused her to knock the little girl's hand away from her mother's. The mother got out, she noticed, but the little girl did not. She herself then lost consciousness and collapsed in the doorway, but the momentum of the panicking crown behind her happened to push her onto the wing and to safety.

Another couple who had escaped from the burning plane just in time by pulling themselves up through a gap in the ruptured roof, described the lasting trauma of the horror that had stayed with them for the four years since their ordeal. Two of their friends had died in the plane. In a BBC radio interview one of these survivors, a young man, told how he now cried more easily and suffered from feelings of guilt because he felt he could have done more to help other passengers on the day. Both of them lost faith in the safety of aeroplanes and took care to avoid them until, a year after the accident, the airline offered survivors a free flight to restore their confidence. The couple accepted but when they got to the airport, the apprehension brought on an attack of vomiting. He said that when they boarded the aircraft, he started to hyperventilate and they both felt so claustrophobic that they abandoned the flight. They declared that they 'had never been the same again' since the day of the disaster.

Inevitably, at the coroner's inquest the recriminations began to flow, although on the credit side the evidence was to galvanize the authorities into action. As will be seen, some more profound changes for the future had been shaping up during the three years and more that passed before the publication in March 1989 of the Air Accident Investigation Branch's final report of its investigations. When the report eventually appeared, it was largely confirmatory because most of its content had already leaked out into the industry and the causes of the accident had declared themselves.

Certainly there were lapses but, taken as a whole, Manchester simply replayed the sequence of events that are well enough known to follow an airliner fire. Most of the actors involved did their damnedest to prevent the tragedy and most of the blame lies further back among the unlearnt lessons of the past. In only one respect was the British Airtours B 737 fire atypical. The airline's emergency drill required pilots to turn off the runway to free it for following traffic, and this had an unfortunate effect. Ideally, planes take off into the direction of the wind, and by turning sideways onto the taxi-way, the crosswind now coming from the left fanned and bent the jet-like flame from the left engine onto the back half of the fuselage. The wall of the cabin was consumed in seconds.

But the chance of such an unusually rapid fire recurring from this cause is far less than that of repeating the classic fate of those inside who die from asphyxia – as indeed did all the Manchester victims, notwithstanding the accelerated effects of the fire. Some have insisted that the main lessons to learn from Manchester are, first, the avoidance of engine breakdowns in the first place, and second, that fire-stricken planes should stop on the runway facing into wind. (More technically, they also rightly counsel a change of priorities in existing crew crash-drills of the many 'vital actions' to ensure that both fuel supply and engines are cut immediately.) They point to one

191

similar event at London Heathrow in 1968 when a BOAC Boeing 707 landed in flames after an engine fire. The crew omitted to cut off the fuel and a severed pipe fed the flames. Five people died and the fuselage of the fire-wrecked plane bore a striking resemblance to that of the Boeing 737 at Manchester.

But it seems undeniable that by far the highest common factor in the record of crash-fires is death from asphyxia – in other words from burning foam and plastics. The causes of aircrashes and the fires that often follow them are legion and, except in an ideal world, the mistakes that create them will continue to be made. So the idea that we should stop crashes happening in the first place must be a counsel of perfection that can never be fulfilled. The proximate cause of *death* presents a single target for action and puts them within reach of a narrower range of remedies. Apart from the airlines and their insurers we are, after all, not much concerned with broken aeroplanes and twisted metal so long as nobody is hurt inside them.

Medical experts pronounced Manchester a survivable accident, so that neither the exploding engine, the fuel drills nor the crosswind and other misfortunes need have killed the occupants. Without the toxic fumes – according to the verdict – they could have survived. Certainly the common causes between the London and Manchester accidents should not be pushed out of the picture, but they cannot claim first concern.

British Airways is reported to have had an 'epidemic' of problems with the engines on the B 737 fleet, none of which had been thought to be a serious threat to safety. Not long before the crash their US manufacturers, Pratt and Whitney, had sent a warning telex to users, but it was claimed that the airline engineers at Manchester had not given it the priority intended. Like other models of the engine, the offending unit had been giving trouble to aircrews. Its suspect behaviour had been logged by the pilots the day before and engineers had given it a normal check-over before the fatal flight. Whatever the strength of the warnings, nobody expected the catastrophic crack in its combustion chamber to cause such a lethal explosion and rupture a main fuel pipe at the most perilous moment of the take-off.

But there were other mishaps. When the firemen reached the plane they found that the nearest two water hydrants were dry – someone had turned them off, and precious seconds were lost in reaching a third. In her eagerness, a stewardess had opened a rear starboard door and deployed the escape chute before the plane had come to a halt. As it turned off the runway, the crosswind blew the flames under its belly and destroyed the chute before it could be used. In the event, only two out of the six emergency exists could be used. All three on the port side faced the flames, and the rear starboard door now lacked a chute. That left only the starboard front and wing exits. Those like Royston Metcalf, who made for one of those two, were the lucky ones. Yet the Boeing 737 is certificated to fly with six emergency exits.

However foolproof safeguards may seem to be, their designers cannot predict every caprice of fate and, to paraphrase the Scottish bard, the best-laid plans of mice and men will ever go astray. Whatever lessons there may be to glean from the human actions that took place on that day, they should not be allowed to distract us from the pursuit of the real killer that is still at large – burning plastics. The failure of the industry to remove or disarm the culprits has a long and discreditable history.

But it would be a mistake to see the Manchester lesson as the only spur to official action. It could have come after any of the 74 airliner fire disasters so recorded over the last two decades, many of which bear a sobering similarity to Manchester. A few paragraphs I first wrote for the *Sunday Times* described the foam-and-plastic menace in the wake of the Saudi Lockheed 1-11 fire disaster at Riyadh in 1980 (see pp. 1–11). It was our first audible bleep – a whisper, if not a call to action – about a danger that figured time and time again in the years that were to follow. Newspaper readers (and editors) have short memories and the same message was delivered with a depressing frequency as each successive fire disaster struck the headlines.

Several manufacturers of fire-resistant foams and textiles brought their wares in to the office, telling tales of unresponsive airline buyers. Some of these products were either costly or had potential snags – others did not. One product we tested over a Bunsen burner on the newsdesk and photographed its evident refusal to burn. Meanwhile aviation authorities in Britain and the USA spanned the years with grave pronouncements couched in committee-jargon, urging the need for caution, a review of wider fire safety perspectives and ever more profound technical analyses. What they were trying not to say was that the airlines jibbed at the cost of safer materials (a little more weight carried on every flight spells a lot more fuel).

Similar events were happening on the home front. Fire brigade officers have known and said for long enough that countless household deaths have been caused by toxic fumes from burning foam-filled furniture. For more than a generation the unlearnt lesson was obvious enough from the number of house and office fire deaths so regularly reported in the press. It was only after a particularly gruesome series of fires in which ten infants died in the first week of 1988 from the results of burning foam furniture that the firemen rebelled in fury at government inaction to such effect that it caught the notice of the media. At last officialdom was jolted into promises of stricter safety standards for household foams which were finally implemented in November 1988.

It is a measure of the media's headlong concern for the moment (and surely a disregard for its own news-files) that a *Times* leader-writer felt himself able to declare that 'so prompt a response to public concern is gratifying'. The memory of administrators was equally short. The day before the leader appeared, the government minister responsible had pronounced that 'there was no point in rushing new controls'. Four months after these words were spoken, as it so happens on the day of writing, the morning paper reports two household fires in which eight lives were lost, six of them children. Both, we are now told, were caused by foam-filled furniture. That death toll began in the 1950s.

The firemen's protest over household foams found allies in the airport and local Manchester fire brigades who had given evidence at the inquest into the 1985 plane disaster. It was now imperative, they said, that the airlines be compelled to deal with the dangers that had caused such a needless death toll in the past. Now *The Times* boldly declared that 'tough new regulations aimed at preventing aircraft fires, including a ban on flammable plastics, are being introduced by the aviation authorities'.

In Britain the CAA had quickly announced a package of measures. It had already imposed a higher standard for fire-resistant seat covers for both new and old airliners on 1 July 1987, six months ahead of similar action by the FAA in America and to be followed by other major countries. Now safer cabin wall and ceiling materials would also have to be fitted to new aircraft. Other fringe improvements included floor-level emergency lighting to help escape from smoke-filled cabins and improved access to over-wing exits.

But the provision of smoke-hoods for passengers, widely canvassed as the obvious solution, was rejected by the CAA at least for the time being. There is in fact a strong case both for and against their use. The best of them might allow people to breathe safely for up to 10 or even 30 minutes and so possibly extend life in an in-flight cabin fire long enough to land the plane (the fate of the 23 victims in the in-flight fire aboard the Air Canada DC9 in 1983 comes to mind – see pp. 185–6). Undoubtedly – if properly used – they could hugely improve the chances of escape in post-crash fires.

One problem is how to teach people to put them on and operate them correctly. Is it feasible to convey this to several hundreds of people before flight when, apart from children and the elderly, incorrect use can be fatal? In some types this can lead to a lethal build-up of exhaled carbon dioxide, and others use potentially hazardous chemicals to filter the air. Hoods are bulky, difficult to stow without becoming damaged – and, despite this, said to be easily stolen by the light-fingered.

To avoid these and other pitfalls, the CAA set a high technical standard which, it finds, no manufacturer has yet met although the search for the ideal hood continues. Those who have read all the pros and cons will probably agree that the case for hoods is as yet unproven. Events have in any case ensured that they are no longer a first priority. A clue to this change comes at the end of the CAA chairman's recent statement about research into smoke-hoods. 'We are vastly encouraged,' he said, 'by the recent development of cabin water sprays' – a development that provides the dénouement in the next chapter.

15 | Saving our skins

Fire-fearing passengers seem about to have their prayers answered at last, not by the removal of all the dreaded foam seats and cabin plastic, but by a method of rendering them innocuous in a fire for long enough for passengers to escape from the plane unharmed – apart from the risk of a slight wetting about the person.

What is particularly significant is that the technical remedy arrived by a side-wind and not through any initiatives by the authorities or the airlines (in fact rather the reverse, as will be seen). It is also interesting to note that the solution is an essentially simple one that had been long ago rejected by officials as naive. A retired fire officer, writing to the magazine *Flight International* in March 1988, tells of his involvement in a similar project fifteen years ago and reports: 'I was unable to motivate any interest in it.'

The media have already extolled the virtues of an 'invention' that simply sprays water droplets from the ceiling and sides of a burning cabin. After the results of successful trials unveiled in March 1988, what remains to be seen is how long it will take before every airliner will have to install it. Detailed evaluations are needed before the industry gives it its final seal of approval, but the more that is publicly known about this potentially vast improvement in cabin fire protection, the more of us there will be to bark at the heels of the legislators and quicken their pace towards the day when it becomes a compulsory piece of safety equipment in every airliner.

The estimated cost of installing the system on a medium-sized airliner is £250,000 or about $\frac{1}{2}$ to 1 per cent of the value of the plane. At first sight the system seems amazingly obvious as no more than a simple adaptation of conventional anti-fire water sprays found in stores and offices. Its ingenuity lies deeper, and a hidden spot of magic lies at the heart of the invention. Schoolroom science is enough, however, to grasp how its first principles are applied.

When you get out of your bath or shower you feel the cold because the film of water on your body is absorbing heat from you as it evaporates into the air. This physical process (known as the latent heat of evaporation) means that wet surfaces constantly cool themselves until dry. The second principle is more esoteric. If the size of a water-droplet is halved, for a given volume of water the surface area is doubled (the calculation is not as easy as it may appear). Very tiny droplets therefore have a miraculously enlarged cooling effect, since the rate of cooling is proportional to the surface area of the water exposed to the air. But this is not the whole story, and at this point we topple into complexities beyond lay understanding because, we are told, the optimum cooling and quenching effects – to deal with both humans and materials – demand a mixture of different-sized droplets. So the need was to design a nozzle capable of producing the necessary mixture.

It came as the brain-child of Jim Steel, a former coalminer who had involved himself in safety equipment and who also became a private pilot. As a safety man, he saw that 85 per cent of aircrashes happened near airports and that the number of deaths from fire was startlingly large. Then came the Manchester airport disaster. His two interests merged when he fell in with Ray Whitfield, a naval scientist and one-time managing director of Rolls-Royce at Bristol. Steel's invention led to a partnership under the name of SAVE.

At first the authorities were not receptive to the idea, and a good deal of time was lost before the upper echelons at the Civil Aviation Authority could be courted. The military were more obliging. It was the enthusiasm and help of the Royal Air Force that gave the partners their first break in the summer of 1987, with the offer of an old VC-10 airliner hulk in which to conduct the first trials. The resulting success of the system allowed them to return to the doubters at the CAA and tell them that it worked. 'There were indeed some red faces there,' an emissary from the company recalls. 'They were under political pressure to satisfy the smoke-hood lobby, so that in a way the advent of SAVE got them out from under.' But the going was still hard.

Then in November came a television slot on the BBC's *Tomorrow's World* science feature that had screened the trials and showed passengers walking out of a blazing cabin unharmed. 'Since then,' says SAVE, 'we never looked back once the show was seen by the public.' More trials were staged for observers from the airlines and

197

the CAA, joined by FAA and Boeing experts from the USA. Some were thoughtful but all were impressed.

The trials demonstrated that the ingeniously metered spray of water-mist stopped fires from propagating in the cabin and extended the escape time (now set at 90 seconds) to at least three or four minutes. And more, as will be seen. SAVE had set out a sequence that simulated a typical post-crash fire in which a pool of fuel ignites under the wing and burns through into the fuselage within 10 to 30 seconds. Then comes a familiar chain of events. Thirty seconds later the cabin fills with black smoke and toxic fumes as the temperature rises – highest near the roof. Then the solid particles in the smoke burn explosively, causing a 'flash-over' fire along the ceiling that engulfs the whole cabin in seconds. Oxygen levels fall as fast as the toxic gas is emitted until the cabin environment becomes unsurvivable. The result was rather different after the SAVE system had been installed.

Its layout is simple. In a typical 150-seat plane it consists of three pipes running the length of the cabin – one along the ceiling and one along each side. Four lateral hoops of piping circle the interior of the cabin at intervals and connect up with the other pipes overhead. The water-misting nozzles are fed under pressure from a 40-gallon inboard tank that is enough to fight the first three to four critical minutes of a fire. Independent cut-off valves ensure that if a section of the piping is damaged in a crash, the remaining network continues to function.

Meanwhile, when the airport firemen reach the scene, they can plug in to any of six external feed valves to keep the system flowing indefinitely.

The trials used human guinea-pigs to endure a fire to be lit around the cabin (one of the volunteers was a brave survivor of the Manchester airport disaster). A 200-gallon tray of fuel was lit outside under the tail of an old BAC Trident and reached a temperature of 1,400 degrees Celsius without even burning through the cabin skin, due to the cooling effect of the water-misting on its inside surface. The occupants reported no ill effects and found they could breathe, see and move around normally. The sprays had been triggered 40 seconds after starting the fuel fire outside. Another test quenched blazing gasoline-soaked chairs in three seconds and cleared the air of smoke and gas.

In a third and fiercer test, the water-misting kept the cabin temperature down to a tolerable 30 degrees C, and after three and a half minutes the inmates daringly switched off the misting. Black smoke blasted back into the cabin so fast that, as one said, it almost beat them to the escape exit. The test team, who wore no protective clothing, switched on the sprays again and the smoke soon disappeared, leaving 'a breathable white fog'. The foam-filled seats that had been exposed to the fire showed that only the corners touching the flames were damaged. There had been no chain reaction to adjacent seats. The only minor mishap of the SAVE trials was that when the smoke had cleared, three of Jim Steel's magic nozzles were found to be missing – perhaps a flattering tribute to his secret.

One extended use of the system has yet to be evaluated – how far it can deal with in-flight fires safely and without threatening the plane's vital electrical and electronic gear. The advent of new 'fly-by-wire' total control systems, which essentially translate the pilot's manual control commands into computer-aided electronic impulses, calls for more research to test the effects of a deliberate soaking. At present it is thought that a captain with an uncontrollable cabin fire on board might *in extremis* accept that risk rather than a worse fate. Fortunately in-flight fires are rare compared to the number of post-crash ground fires. The prospect of inadvertent triggering of the mechanism in flight does not seem so critical, as damage from the spray would take considerable time. It would provide its own early warning – not least to the passengers themselves.

Watchful air travellers will hope that the benefits of SAVE will come sooner rather than later, and as soon as the authorities can finally endorse the claim of its chairman, Ray Whitfield, that the water-misting trials 'conclusively prove beyond any doubt that the system will keep the fiercest fires from entering the cabin, giving the occupants time to escape'.

Has the foam and plastics menace then been overcome, and is there any need to press for safer substitutes? If water-misting can keep the cabin temperature down to a safe 30 degrees C while a 1,400 degrees C fire burns around it, neither the plastics nor the occupants would seem to need further protection. But this assumes that SAVE has not only emerged fully proven from the experimental stage, but also that it is competently installed throughout world airline fleets. Experience warns that this happy state of affairs could be years ahead. Secondly, every system has its bugs and failures, and all of us would sit a good deal happier in our seats in the knowledge that they themselves were fire-proofed with the new materials now available. Both the CAA and the FAA have wisely adopted a multi-pronged attack on cabin fire dangers, and it is to be hoped that they will not slacken the pace towards banishing polyurethane foam and flammable plastics from passenger cabins. The message to airlines should be, cut the luxuries a little and get rid of the hot seats now. It is reassuring to know that among other airlines well along this road are British Airways and Lufthansa.

It is worth keeping in mind that the inquest into the Manchester airport deaths pronounced it to have been a survivable accident. It may be idle to hypothesize, as some have done, about the number of lives that might have been saved if Jim Steel's invention had been installed in the plane before that fire. But his name may at least go down in history as the man with an idea that *could* have saved a thousand lives, but for a timid and cost-fearing industry.

The untimely death of Jim Steel in July 1988 has not been allowed to hinder the exploitation of his invention. Joint evaluation trials between the UK Civil Aviation Authority and the US Federal Aviation Administration began in May 1989 at the FAA's Technical Centre in Atlantic City. The British system has been installed in an old DC-7 fuselage and in a jumbo DC-10 mock-up, while the CAA has also commissioned its own water-mist tests on a B 707 fuselage at the UK's Fire Research Station at Cardington, England. And with Canadian participation, Airbus and Boeing have been contracted to study the engineering and airworthiness aspects.

While thorough testing is essential, the pace of the regulatory authorities thereafter seems decidedly leisurely: it is reported that their collective aim is to reach a 'mid-1990 decision' on whether water-mist systems should become a standard airline safety requirement. Having reached that milestone, and judged on past form, it could be as much as three years before the hardwear is manufactured and fitted to all newly-built airliners as mandatory safety equipment, and far longer before the airlines could be required to retrofit their existing fleets.

Assuming that the current rate of fire fatalities continues, many lives will be lost during the ten years or more before passengers are given this remarkable protection from the worst hazard of flying. As soon as the necessary hardwear is on the shelf, we should be demanding faster action from the administrators. One of those who will be doing so is John Beardmore, a survivor from the B 737 Manchester airport fire disaster, who has formed an action group* with other survivors of Manchester and other recent disasters, with the aim of speeding up the action. The survivors, he says, are shocked by the delays and the little that has been actually done to remedy the dangers that trapped and killed so many people before their eyes inside the burning inferno at Manchester. 'If it happened again', he believes, 'it would bring the same result.'

Apart from some shining exceptions, airlines are not known for their eagerness to speed the day towards the extra cost of mandatory safety measures. On the basis of current costings, the installation of the system on a new jumbo is of the order of half of one per cent of its value. It would be interesting to see the results of a poll of airline passengers, who might be asked whether they would be ready to pay a proportionate increase on the price of their tickets to buy this insurance against disaster.

* Survivors Campaign to Improve Safety in Airline Flight Equipment (SCISAFE).

16 | The sky is full

Although the fears of the passengers who were ringside spectators in a number of recent series of airmisses cannot be dismissed, the avoidance of mid-air collisions is one branch of air safety that is least amenable to positive improvement, given the conditions in which we allow the industry to grow to bursting point. In the lower traffic densities of the less crowded skies of yesteryear, a glance at the safety record shows that airmisses were more frequent than they are today. In other words, a much higher standard of safety has been achieved, despite the doubling of traffic over the last decade. This is a broad figure, but the overall trend is clear: it looks as though we are doing pretty well. Is it complacent to expect this improvement to continue?

The first difficulty to face is that there is a lack of consensus about the current rate of airmisses. The policemen of the skies, the regulatory authorities, have a vested interest in playing down the risks. The unhappy pilots and air traffic controllers – who are virtually the only potential offenders in the act – are bound to stress the operational and environmental hazards in their own defence. Behind the front line, the inevitable clash of sectional interests continues out of the public eye as government accountability on the one hand faces industrial and union claims on the other. By the nature of their being, governments find it convenient to tag their opponents with a political label and, so far as air safety is concerned, it is regrettable that this only serves to muddy the waters in a way that makes it all the harder to identify the areas where further improvements might be gained.

These back-stage contentions have polarized ever since President Reagan fired his 11,400 striking controllers in 1981 – an action that seemed at first to be vindicated by a drop in all types of fatal accidents suffered by US airlines. Comparing the four years before and after 1980, they fell by 34 per cent and killed 74 per cent fewer people. By 1984 the death toll had shrunk to only 54.

But in the following year it had leapt up to 561. Meanwhile mid-air near collisions jumped from 589 in 1984 to 777 in 1986, prompting the US General Accounting Office to warn the Transportation Secretary that 'the present system does not provide the same level of safety as before the August 1981 strike'. Expectedly, the Reagan administration denied any link between the wholesale firings and the rising totals of collisions and near-misses, preferring to point to other causative factors including the increase in traffic and the competitive effects of the deregulation of the airlines.

On 13 February 1989 two planes nearly collided above Seal Beach, California. Both had been directed to fly at 9,000 feet and on courses which were converging. Due to the rapid action of one pilot, they missed each other by a horizontal distance of 1.9 miles. Air traffic controllers had made an error. Although serious, it was not exactly an uncommon event, but the inquiry into it revealed a history that should disturb us all.

To set the scene, a British Airways Boeing 747 jumbo with 286 people on board had climbed out from Los Angeles international airport a few minutes before en route for London. Flying at the same height and on a converging course was an American Airlines British Aerospace 146 with 70 people aboard it on a short-haul route to Ontario international airport, also in California. The British pilot, warned at the last minute of an approaching collision by the controllers, managed to turn and climb the jumbo just in time to avoid it.

The National Transportation Safety Board immediately examined the causes of the mistake. The seriousness and magnitude of the deficiencies in air traffic control, it said, prompted an extensive investigation. This followed the discovery that eight operational errors – that is, breaches of the minimum safety separation limits between planes – had been committed by the air traffic control centre concerned in less than a year.

Inadequate controller staffing and excessive overtime work were among the chief causes of the mistake, they reported, and these warranted immediate attention from the Federal Aviation Administration. Documents revealed that, following the controllers' strike in August 1981, the FAA had reduced the level of authorized controllers from 48 to 30.

By January 1989, there were 51 staff while the current authorized level stood at 66. The Safety Board found that under-staffing, excessive overtime and deplorable working conditions had existed for several years and had been fully documented to the FAA, but that the Administration had failed to address the problems. Part of one of the NTSB's recommendations to the FAA is both surprising and revealing. It reads:

> Conduct a staffing study to determine if aviation-oriented persons from the local area, such as retired pilots and military personnel, could be hired at the Coast Radar Approach Control to perform the duties of air traffic assistants.

We are not told, however, whether or not any of the controllers sacked by the President in 1981 were to be regarded as eligible for these stand-in jobs.

Britain mirrors a less dramatic controversy, but it leaves its airline customers just as perplexed as their American counterparts. With air traffic presumed to more than double by the end of the century, the authorities are already battling with the congestion in the skies in which there have been so many recent near-collisions. Yet, as the search for remedies continues, a union spokesman for air traffic controllers claimed in July 1989 that they were handling 40 per cent more traffic since 1983 with 20 per cent less staff. They believe that things are worse than they seem.

British controllers claim that many airmisses go unreported so that the official figures are invalid. Before 1988, when the rules were changed, only pilots could file official reports of an airmiss. If it has been caused by an evident lapse by the controller, the pilot may be unwilling to get his colleague on the ground into trouble and so forget the incident. The same fraternalism between pilot and pilot may let others go unreported.

There is some evidence for this in the pages of CHIRP (pp. 41–2), the anonymous confessional that is circulated internally to pilots and controllers. But if the official figures have been invalid for these reasons, part of the remedy would have been to allow controllers the right to file reports, a right that they have only recently won.

The suggestion that collusion between pilots and controllers hides other near-misses must – apart from CHIRP – remain largely a suggestion because, whether they are few or many, there is a high chance they will not be detected. (A rare exception to this came after the much-publicized close shave over the Atlantic in 1987 (p. 000): other aircrew who were listening on the same radio channel heard and recorded the pilots proposing such a secret deal.) So until somebody can present us with a better method of chalking up the real score, any disputes about it are rather sterile. With the caveat that it may be rash to draw any firm conclusions, the official airmiss record remains as the only yardstick to hand at the moment.

But it can be said that fewer people are being killed in mid-air collisions than in the past, despite the huge rise in the traffic that is expected to double again well before the end of the century. Will the nexus between pilot and controller be strained to breaking point before this comes about?

Air traffic control

The working environment in which these navigational mishaps occur is a strange and solitary one. The figures that can be seen watching from their tinted glass eyries in airport towers are there to handle arrivals, departures and taxiing on the ground. Outside the airport zone, constant radio contact with pilots over the entire duration of each flight is maintained by remotely sited air traffic control centres. Here in dimly lit enclaves controllers sit ranged behind a bank of video screens which show the moving pattern of each plane as a fluorescent spot or radar 'blip'. The general hush is broken only by the rattle of terse radio talk between each pilot and controller. The supervisor may be leaning over one sector controller's desk to monitor a busy burst of traffic on his screen or check the 'hand-overs' from one sector to another. How all this builds up into a coherent picture of the real airspace outside calls for a brief recapitulation of the way that the skies are regulated (such an outline calls for the indulgence of any professionals who may notice it).

Airspace is divided into controlled and uncontrolled segments and we are concerned mainly with the former. This contains the 10-mile-wide commercial airways – the motorways of the air – that are set in the skies at various levels between about 10 and 40 thousand feet, and the terminal areas around airports. Controlled airspace can be seen as three-dimensional chunks, like layers of cake poised at different levels. The base of these chunks lowers towards the terminal areas until the last of them reaches from ground level up to a fixed safety height (often between 3 and 6 thousand feet). All the rest is uncontrolled airspace mostly inhabited by light aircraft. (Not always: airlines can enter it and many airmisses occur in this shared airspace which is also used by military traffic.)

Both the airplane and the pilot must be equipped to fly in controlled airspace. The plane must carry a full complement of radio and navigational gear and its pilot must hold an instrument rating – for the reasons that follow. Such flights must be conducted according to Instrument Flight Rules (IFR) which means that flight is by instruments and not by eyeball, in other words blind-flying techniques are employed both by day and night (apart from the landing, one hopes, although even this can now be nearly totally automated).

The point that concerns us most, of course, is that under the almost wholly pervasive IFR operations, it is the controller sitting at his radar screen, and not the pilot, who is effectively responsible for navigating the plane far above him and for avoiding collisions with other planes under his control (although the pilot has an overriding veto). The fundamental rule is that flight paths must be separated from each other by a horizontal distance of 5 nautical miles or by 1,000 feet vertically (the exceptions need not concern us).

The normal routine before a flight is for the pilot or his airline to file an IFR flight plan, usually in a standard form, setting out his route and the estimated times along it to the destination. This is then magically flashed to all the ground control centres affected.

When the aircrew clamber aboard and turn on the switches, both they and air traffic control (ATC for short) have the flight plan details before them, so that the forthcoming radio contacts between them are minimized to a brisk and virtually coded exchange of orders and responses. It must be so, because several other flights will be sharing the same radio channel. When the plane is ready for departure, ATC reads out the pilot's clearance to enter the route set out in the flight plan, subject to changes of times and other variables.

Once airborne and up in his allotted airway, the pilot reports his arrival at each successive waypoint on his chart to ATC: time, flight level, and his estimate for the next one. These waypoints are either ground-based radio beacons or a point fixed by a set distance from them. Nowadays there is a back-up to this formal ritual in the shape of a little tell-tale transmitter on the plane, the transponder, that emits automatic signals to ATC radar screens and confirms the plane's position, flight level and identity.

The scene is now set for our airmiss. It doesn't take much explaining. The pilot can suddenly make a mistake – a turn, climb or descent contrary to plan – or ATC may give him a wrong, misleading or ambiguous order. Anyone who has ridden a tandem bicycle will recognize the scene ('I said *left*, not *right* – Oops'). The danger is of course that in a dark and cloudy sky there are a lot of other planes to bump into, and a lapse of a few seconds can invade the allotted flight path of another.

As might be expected, there is a labyrinth of technical complexities that weave themselves around such introductory remarks about the workings of air traffic control. But the elements that concern us are pretty near the surface.

Other mishaps can occur. ATC's computers that feed the radar screen may suddenly go on the blink or radio contact between plane and ground may be temporarily lost (a recent airmiss between Concorde and another plane near London, for instance, was caused by a pilot's radio transmit-button sticking, with the result that the radio channel became jammed during the vital seconds of an avoiding manoeuvre). But such failures are rare. All but a few airmisses or collisions are caused primarily by a human error on the part of the pilot or ATC – by man and not machine.

It is for this reason that the analysis of airmiss scares seldom tells us much that we don't already know – that it was caused by a misunderstanding or hiccup in the communication between pilot and controller. It happens for equally simple reasons.

The truth is that the method of communication is both primitive and slow compared to the rest of aviation's hi-tech image. Vital data must be translated into fallible speech and then transmitted by radio to an equally fallible ear – data of a kind that so palpably lends itself to other existing methods of direct contact between the electronic systems on the ground and in the air. The closest paradigm here is the airliner's transponder that periodically bleeps its progress direct to ground computers by transmitting encoded radio signals. Again, pre-recorded flight navigation tapes can already be fed into the flight management systems that control the plane so that most of the route can be flown on autopilot – even the initial landing. Here the pilot's role is no longer to manhandle so much as to monitor the flight.

Technological innovations

Consider the modern marvel of inertial navigational systems (INS) that are nowadays carried on most airliners, and which are capable of pinpointing their position with an accuracy of a few yards anywhere on the globe. Pre-programming allows them to guide the plane to each successive waypoint over the entire route. The two or three INS sets that are normally carried interlock and countercheck each other so that the chances of an error or total failure become astronomically small. When the INS outputs are fed into the flight management system, the whole becomes a surrogate pilot. The live pilot need then only monitor the navigation. INS began as a back-up to conventional navigation from beacon to beacon, but it is now a primary tool on long-haul routes.

But these miracles of precision in space and time cannot predict the presence of other planes or the onset of a coming collision. This crucial need still relies on word-of-mouth warnings and responses over busy radio channels (and often, as we have seen, in halting pidgin English) – a proceeding that seems light-years behind most of the other sophisticated technology that is now taken for granted elsewhere in aviation, not to mention flight in space.

For the sake of simplicity, this view has been somewhat over-stated. Collision avoidance devices have been discussed for years and some are now in the final stage of development. The US House of Representatives has called for such a system to become mandatory on all US fleets by November 1991. But the snag is that this cockpit collision warning kit – called Tcas II – is still under trial and it is uncertain whether it can be proved within this time-frame. Fears remain that this method of avoidance between pilot and pilot – unaware of the whole air traffic pattern in a busy sector – might steer them into other traffic. Technology may eventually iron out this and other doubts, and at least a start has been made.

Meanwhile Britain has taken a different approach by testing a system that will warn the air traffic controller rather than the pilot of a potential collision. Known as Conflict Alert, it is designed to predict the forward courses of all the radar blips transmitted by each plane that show on the controller's screen and flashes an emergency warning if a collision is imminent. So far this has been left to the good eyesight and watchfulness of the controller himself, who may have to monitor up to a dozen blips moving on his screen.

208

Inevitably there have been mistakes and public alarm after a near-miss over Kent close to the English Channel – between a British Airways TriStar and a Bulgarian Tu 154 charter jet in February 1988 – hastened the first installation of Conflict Alert at London's main ATC centre at West Drayton near Heathrow. But the system must still rely on word-of-mouth commands by radio from controller to pilot, with all its imperfections. Unlike the American system, it is the controller who is left with the task of choosing the safest avoiding action and which plane to direct first. To a lesser extent, the problem remains of one avoiding manoeuvre-creating threats to other traffic.

Under either system, the controller has a daunting responsibility, and at fraught moments – now becoming more and more frequent – the skills needed are immensely demanding. Against the habitual and often misconceived press criticisms of the UK air traffic controllers, it needs to be said that the British service has never seen an airline passenger killed in a collision. Homage has already been paid to the high ATC standards achieved in the UK, USA and elsewhere. They cannot be said to be universal, and the international traveller still has to run the gauntlet through too many poorly controlled skies. The western examples offer the best hope of improvement in those regions known to pilots as 'Indian country' (Red Indian of course).

So it appears at the moment that the weakest link in the chain of collision avoidance strategy is the swapping of words over two-way radio sets and that this is likely to remain so for some time to come. The fallibility of that primitive system – a not-so-distant legacy from the era of flag-waving, semaphore and Morse code – is a pretty direct measure of the extent of the collisions and airmisses that will continue to afflict us until it is replaced by an omnipotent substitute drawn from the armoury of electronic and digital know-how of the kind now deployed in the other systems of airliners. The real 'pilot' may then be sitting comfortably in a simulated flightdeck on the ground transmitting his commands directly to the airplane above, with only a monitoring team on board. Futurology apart, until such ideas are fully tried and tested, our instincts may prefer that we stay with the present arrangement and fly in company with a live pilot, despite that method's evident imperfections.

Deregulation and congestion

But such technicalities distract from the fundamental cause, if not of every airmiss, of most of those that are reported. In the wake of intense alarm at the near mid-air disaster to President Reagan's plane in August 1987, it was disclosed that near-misses involving US commercial flights are occurring at a rate of more than one a day (and three a day if all types of plane are included). The policy of deregulation there – opening the skies and airports to anyone who shows that he can run an airline – has certainly brought a measure of the chaos to air travel that is so much in the news, but this should reinforce rather than obscure the universal fact that airmisses are a function of the number of planes in the sky.

On the other side of the pond, it was this realization that caused a voice at Brussels to make a plaintive pronouncement that is too seldom heard in official circles. It came from Karl-Heinz Neumeister speaking for the Association of European Airlines at the Transport Commission hearings. 'The sky,' he concluded 'is full.' Who has allowed it to become so?

It has been said that aviation is the only industry that is free to grow without limit or constraint, a notion that has become an established article of faith shared by both politicians and operators. Those who dare to speak it aloud tend to be regarded as faintly indecent, if not heretical. Yet unrestrained free enterprise and safety have never been happy bed-fellows.

In a recent analysis of the results of deregulation an American pilot, Captain Roger Hall of the Airline Pilots' Association, complains of 'the reduction in safety margins due to carriers squeezing to reduce costs' in a context of fierce competition and over-supply. He cites an increase of 44.3 per cent in the number of weekly departures from US hub airports in one year to June 1987, and an increase of 54.6 per cent in the seats available on them. 'A philosophy of laissez faire towards air transportation,' he concludes, 'will in the long run weaken a country economically.' Meanwhile he sees safety, the travelling consumer and the employees as its first victims.

As a generality, pilots are a pretty conservative body of men, and it would be a mistake indeed to regard this as a political stance. The buck stops with them. Whatever views they may hold must be tempered by the truth that it is, after all, only the pilots and their passengers whose lives are at the greatest risk from air collisions.

Those who declare that the skies are less safe than they were face the inherent difficulty of demonstrating the part played by commercial imperatives as a cause in any particular accident. Take, as one instance among many, a fatal crash that happened in Europe even before the advent of de-regulation in the United States – the world's worst air disaster at the time.

Two Boeing 747 jumbo jets collided on the fog-bound runway at Tenerife's Los Rodeos airport in the Canary Islands in March 1977, killing 574 people. A Dutch KLM flight started its take-off run into the fog before a Pan Am plane had taxied clear of the runway. How far was the Dutch captain's precipitate move caused by his need to get off quickly before his crew ran out of duty hours and so lead to costly delays for his company, and how far was it due to a genuine (but mistaken) belief that airport control had authorized him to take off?

The desire to propitiate employers by saving time and money can be an unseen motive behind accidents that show up in the records under other headings such as pilot error, fatigue, stress or perhaps poor communications. What may be said, however, is that there is a consensus among pilots that safety standards are indeed at risk under the fiercer competitive forces unleashed by effects of deregulation as it spreads throughout the industry.

When opinions are based on imponderables, as they are here, they are likely to vary widely. Attempts to back them up with statistics can be unconvincing if their validity is suspect on the grounds that have been suggested. For example, a review by Professor Alfred Kahn – a leading proponent of de-regulation in the USA – suggests that the number of fatal accidents suffered by US airlines in the ten years since de-regulation (1978 to 1988), has dropped by 50 per cent. The accident data-base of the London magazine *Flight International* told its editor-in-chief Michael Ramsden a different story. US airlines, he found, had had 22 fatal accidents in the five years before de-regulation. He compared this with the figures for the last five years (1983 to 1988) when, as he explained, the new regime had settled down, and found a count of 42 comparable fatal crashes – an increase of 90 per cent. Since Ramsden published his findings in January 1988, this apparent contradiction had not so far been resolved.

211

Whatever the true consequences of de-regulation may be, it has put a heavier burden on the regulatory authorities in policing both the sea and air. Seen as a tug-of-war between the safety watch-dogs on one side and the industry on the other, the new breed of buccaneers have brought extra muscle to the corporate team. The Zeebrugge sinking and its aftermath seem to support this analogy. If the thinking of the Royal Institution of Naval Architects is correct, the ships it regards as inherently dangerous will sail on. Both the political and commercial costs of proscribing the design of the ro-ro ferries now in service, are too high. The Department of Transport has imposed minor safety improvements on the fleets, but whatever its own inspectors may feel about it, they must work within the remit set by their political bosses. Safety seems to have lost a good few inches in the tug-of-war with commerce. Realistically, there must always be some trade-off between safety and the commercial viability of an enterprise. But when passengers' lives have been shown to be at stake, one would expect the claims of the market to carry less weight that they may elsewhere in industry. It is not a novel thought.

Meanwhile the Europeans are flirting with the American style of deregulation that can now be seen approaching from the western horizon. The Brussels plan is that after 1992 airlines will, in theory, be allowed to fly wherever they want within the European Community. In Britain, a policy of loosening government restraints and allowing the industry to have its head is well on the march. Its pilots declare that they have always had a profound working respect for the Civil Aviation Authority but some say that it is becoming sadly tarnished.

Captain Ian Frow, a heavyweight among members of the British Airline Pilots' Association, believes that the CAA is no longer the force for good that it once was because it finds itself under new pressures to appease the airlines. The twin dogmas of reduced government interference and reduced public spending are to blame, he says. However effective they may be elsewhere, 'in British civil aviation they are potentially lethal since in some respects they have reduced the CAA to the status of a powerful onlooker in a business that needs strong and effective regulation. It is this weakness which has contributed to the current ATC system alarms.' If this is allowed to continue, he goes on, the travelling public will be scared witless with tales of declining standards. That is your captain speaking.

The hard edge of this controversy rests on the present rate of traffic growth. To suggest that the best way of stopping planes from knocking into each other might be to have fewer planes up there in the first place is to break the industrial taboo that was mentioned earlier. There has been one notable and creditable exception to this. In October 1987, coming under pressure about a series of airmisses and ATC breakdowns caused by the rising workload, Christopher Tugendhat, the chairman of the CAA, frankly declared that 'ultimately the volume of this [traffic] will have to be capped . . . although the timing of this will, of course, depend on how quickly technology evolves'.

That was a brave message indeed, considering that his Authority not only operates the air traffic control system but is also the licensing power that grants approval for all flights to and from UK airspace. If the one cannot manage the other, then both the responsibility and the remedy lie in the same hands. Added to the CAA's duties is the promotion of the aviation industry and all its needs. The chairman's statement (that incidentally went all but unnoticed in the media) is a rare but welcome admission of ultimate dangers that lie ahead.

The warning surfaces again in a more muted fashion in the CAA's annual report published in July 1989. If, despite all other efforts to contain the growth of traffic, it exceeds the capacity to handle it safely, the Authority 'will have to consider ways of restricting access to the system, such as giving priority to scheduled services . . .'. However unpopular restrictions on charter flights may be in the travel trade, they offer the widest scope since it is said that leisure travel now accounts for 80 per cent of all traffic.

Moves towards a unified air traffic control system throughout Europe have been held out, both in Brussels and at Westminster, as the official panacea to the present congestion of contintental traffic. The technical apparatus for an integrated system exists. But progress towards this goal has proceeded at a small's pace chiefly because of the reluctance of politicians to yield up the degree of national sovereignty needed to achieve it. At a European conference of air traffic controllers held in October 1989, speakers saw the daunting prospect of the present delays and congestion continuing 'for 20 more years'. Privately, they believe that the chances of unification in the near future are negligible.

One is therefore drawn to the conclusion that the shrill alarms at the dangers of collisions in the air that periodically sound in the headlines, no matter where the accusatory finger may point, are counter-productive unless they mention the real causes: too many aeroplanes, too much hyping up of the economic benefits from the tourist trade (it's a two-way traffic), and too little regard from the inevitable diminution in the safety of travel that these are even now causing.

Automated anti-collision systems may permit more traffic to fly safely in the future, but until that day comes the choice must lie between a philosophy of unrestrained growth or 'capping' the historic freedoms of the skies, at least to the extent that they are kept as safe for passengers as they have been in the past. Profound inquisitions into the details of each airmiss or collision are a less profitable exercise than a recognition that the industry can't have its omelette without breaking eggs.

The luck factor

The trail of tragic air collisions from the past should forbid anything less than a serious approach, but there is one speculation that may not be as idle as it appears: could it be that luck is the air traveller's best friend?

At present air traffic is artifically confined into aerial motorways and then corralled into spiral 'stacks' where the planes queue up for the order to slide down the single-track approach path in line astern to the airport. Suppose, a little madly, that the whole system was decontrolled so that flights were free to choose their own routes through the vast three-dimensional element of airspace that is now apparently left unoccupied? Leaving aside the inevitable bottlenecks at airports for the moment, some have argued that there would then be a numerically smaller chance of collisions in the skies. Absurd though this concept may seem, a further look at it helps to identify some of the worst knots in the tangle of rules and regulations that we seem to have imposed on ourselves.

The mapping out of the so-called airways, the trunk routes of the skies, has been briefly mentioned earlier and the pattern of traffic management emerges in many of the eventful flights that have been described. These are worth a closer look before questioning a system that looks almost as if it had been designed to herd aircraft together rather than separate them.

By analogy, the airways are the three-dimensional motorways that link the airports of the world. Although they are ten miles broad for safety reasons, pilots must fly precisely along their centre-line. Remember that planes must be separated either by five miles horizontally ('Please keep your distance') or 1,000 feet vertically (2,000 feet at higher flight levels). To follow the analogy, the picture emerges of single-track highways stacked one above the other.

To keep on track on the middle of the airways, pilots follow waypoints marked out by radio beacons sited on the ground exactly under the centre-line. New technology has refined this simple layout somewhat, but its fundamentals remain, and the method of flying from point to point dies hard. It follows that all planes travel like tramcars along predetermined grooves, ignoring the volumes of empty airspace all around.

There are surprisingly few of these of airways and there are, for instance, only some dozen such major routes radiating from London. With a dash of romance, each one rejoices in its colour-coded designation – Green One will take you over Strumble in rocky Wales to Shannon and the Emerald Isle and thence to the Big Apple, Blue Four invites you to the shivering north by way of Cumbria and Glasgow, while gentler Amber One beckons you home again. Red One, with its sinister ring, can carry you eastwards to Holland, Germany and on into Soviet airspace if you will. But doubtless any such fancies are long lost on pilots crawling into their cabins for the umpteenth time in a busy week as they hear the order to fly 'Red One, as cleared . . .'.

So at most hours of the day and night, planes are streaming along these same single-track trajectories. Must they be so confined?

It all started in the days when aerial navigation was a matter of map-spotting points on the ground. This, we are told, included the Baedeker method (from the famous German traveller's guide of the time) – in moments of doubt one swooped low over railway stations to read the notice board. In time, flashing light beacons at night gave way to radio beacons and so to navigation by instruments rather than eyeballs. Like Topsy, the system just growed as the habit of map-hopping became one of beacon-hopping. But then things came to a halt.

Since then technology has served up a whole basket of goodies that now allow pilots to carve out any desired flight path with uncanny

215

accuracy: on-board inertial navigational systems (INS) tell the exact position of a plane to within a few yards anywhere on the globe, distance measuring equipment (DME) gives a constant read-out from a choice of transmitters on the ground, radio altimeters check height as planes descend to land . . . A stream of data from these and other sources flow into the computerized flight management system (FMS) which, when programmed according to taste, can act as a surrogate pilot. It can be fed with pre-programmed navigational tapes that will carry the plane on any desired route with an accuracy not far short of a railway train.

So the potential for discarding the traditional single-track airways is there. Each plane could carve out its own individual route to be approved by a central air traffic computer before take-off (constant data-linking during flight is just round the corner).

Such a concept seemed to border on the realms of futurology when I first wrote about it, but in July 1988 came the news that the professionals were well ahead of me. We are told that the Royal Aerospace Establishment at Bedford, England, is looking at a whole new approach to air traffic management which it describes as four-dimensional (4D) control.

'Despite rapid advances in both ground and aircraft based systems,' says the RAE, 'air traffic control methods have not changed since its inception.' The capacity of UK airspace could be increased by four or five times, it says, by integrating and exploiting more fully all the existing navigational gear now used in the air and on the ground: all that needs to be added is a modified on-board transponder linked to the ground-based collision avoidance system that is now coming into use (Conflict Alert, see pp. 208–9). These, plus a modest change in aircraft software, would open the door to multi-track flight paths, much in the manner which has been suggested earlier.

The 4D concept is so called, the RAE explains, because the plane will be controlled not only in the three dimensions of space as now, but also in the fourth dimension of time. Flight tests are now under way at Bedford, mirrored by similar tests in the USA at NASA Langley base.

But as the pace of technology quickens, the time-lag in implementing change becomes longer, for fairly obvious reasons. World-wide agreement is always slow to come and the costs of the transformation are evident – changing the established route structure, re-training crews, re-equipping fleets and ground apparatus. But the curtain is rising on the second act of this change-over.

The greatest congestion occurs at airports and is worsening yearly, as the recent chaos in the USA and Britain have taught us with a vengeance. Much of this is due to a system that allows only one plane at a time to make its ponderous approach to a single runway. It is safe, but becoming distinctly archaic. The universal instrument landing system (ILS) is little more than a refinement of radio guidance devices developed during the last war. It consists of two fan-like radio beams transmitted from sites on the landing runway, one in the vertical plane (the localizer) and the other nearly horizontal (the glide slope) which meets the runway at a descending angle of 3 degrees. The line where the beams intersect at right angles indicates the safe path of descent over the last six to eight miles to touchdown. The pattern of the vital beams is hard to describe, but one might imagine the pilot as following the centre point formed by the feathers of an over-size dart when seen from behind.

The pointers on his radio navigational instruments guide the pilot to turn left or right, up or down until he captures and holds onto the centre line formed by both beams. If he succeeds (it's an acquired skill), the system will steer him down 'blindfold' out of cloud or darkness until the runway lights come into view.

These are the raw bones of a 'hands-on' ILS approach to land, but nowadays automated flight control systems will do most of the job for the pilot. Breakdowns occur, however, and he must be able to revert to the expertise of a manual approach at any time.

By the time he has landed and taxied off the runway, the next plane in the landing queue some two or three miles back will have begun its final approach pursued by a third settling onto the ILS beams at an equal distance behind. At peak arrival times there may be a dozen others wheeling around in the spiral hold or stack waiting their turn to peel off and slide down the ILS.

This slow single-trajectory approach to a single runway has been with us for generations and, like our traffic control systems, seems to defy the hi-tech ingenuities that are all around us today. Why not allow shorter and multiple-approach paths, perhaps curving in from unused airspace towards two or more landing runways? Such a system has been possible for long enough, and in 1978 by international agreement the change-over was set for the year 1998, but progress has been slow and that target is unlikely to be met for what is to be called the microwave landing system (MLS). Tests in the USA and the UK show that although there are few technical problems, the logistics of re-equipping fleets and airports worldwide are complex and predictably slow. And again, the transition from a system that has been learnt by rote by two generations of aviators is a challenge to human inertia.

As its name implies, MLS uses a much higher frequency radio band that not only brings an abundance of separate radio channels, so that a number of aircraft can fly simultaneous approaches from several directions, but – when linked to on-board auto flight systems – also makes curved rather than straight-in approaches possible. Unlike the single ILS path, the funnel of airspace leading to the runways can then accommodate many more planes. And more than one plane can land at a time. MLS beams are more precise and reliable than ILS, so that big airports with two or more runways can safely allow parallel arrivals to each of them, instead of cramming into a single ILS approach path as they must at present.

If airport congestion has helped the development of MLS, the kind of lateral thinking it employs may inspire the new moves towards opening up the unused skies around the airways by similar techniques. The picture of air traffic as it is today shows vast volumes of unused airspace around and above us and it is here that the provocative theory about luck being the traveller's best friend has a plausible ring. Pocket calculator at hand, it might run as follows.

Imagine packaging a jumbo jet into a box or parcel of airspace of a size that will comfortably fit it from nose to tail and wing-tip to wing-tip. Then calculate how many parcels could be stacked in the immensity of the total airspace available. For example, the whole area of Britain's national flight boundaries and the volume of air above it to a height of 40,000 feet contains about 3 million cubic miles – space enough for 248,899 *million* boxed airliners. According to the Civil Aviation Authority, at peak hours British airspace contains about 400 planes.

Suppose these were freed from the confines of the airways network and notionally scattered at random through the skies like confetti. At any one instant, the odds of any two planes touching each other seem to come out at more than 70 million to one against. The answer to the sum compares oddly with the industry's humbler boast of a million-to-one chance against an accident in the air despite our closely regulated skies.

If all regulations were to be abandoned and planes flew about at random, the luck factor – so the argument goes – should make flying 70 times safer. But at least one flaw in this game of three-dimensional chess in the sky is that it seems to assume that the parcelled planes are not moving, since the odds of a collision are measured at a single instant. Calculating the real odds of a collision over a period of time – and with planes being drawn like a magnet towards airports – seems a task set for a main-frame computer manned by senior wranglers. If this numbers frolic teaches us nothing else, it may extend our conceptions about the vast volumes of unused skies and, like MLS, point to other ways of relieving the congestion in the air that seems at least in part to have been self-inflicted by habit and tradition.

The proponents of the luck theory have at least one valid point. In times past, instrument flying was much less precise than it is today. If, by one of those not uncommon mischances caused by pilot or controller error, two planes were flying identical routes in time and space, there was a good chance that the combined inaccuracies of their radio navigational apparatus would ensure that they would miss each other by a small margin. Nowadays that element of luck has been virtually eliminated by the sheer precision of automated navigation, so that a wrong or misunderstood order leads closer to calamity.

Conclusion

New horizons

There is little doubt that the auguries for safer travel in the air are promising. The course seems set towards larger subsonic airliners carrying up to 1,000 passengers, navigating by safer and cheaper space satellite systems that will at the same time bring the perhaps questionable comforts of telephone, fax and television points to each seat. The precision of guidance and communication via satellite will reduce the chances of collision and delay, as the shortcomings of verbal air traffic control and the queues of 'beacon-crawling' aircraft recede into the past. Pilots will be able to 'see' the outside world from their cockpits with more accuracy and realism than they do today: the displays on their screens will show their position in relation to other traffic and terrain, as well as the airport approaches in all weathers.

Every so often, the technology of flight takes a leap forward on the back of a single new discovery. The high-flying jet engine opened up the skies by doubling both the amount of usable airspace and the speed of travel. The advent of inboard Inertial Navigational Systems (INS – see p. 208) put much more accuracy at the pilot's elbow and so removed the need for flight navigators in the cockpit. Satellite technology is now poised to take us into new dimensions of accuracy, certainty and safety.

Constant links between plane and satellite promise navigational accuracy to within a few metres on any route in the world. This will increase airspace capacity by allowing planes to fly closer to each other without any loss of safety: triangular links between plane, satellite and ground control will eliminate the range limitations of conventional radio contacts and allow the controller to pin-point the position of traffic beyond the reach of radar. The use of digitalized signals increases the quantity and simplicity of a constant flow of information – ground controllers' instructions, weather reports, navigational data and, of course, return contacts from plane to ground. Most of these routine exchanges will be sent automatically by coded signals or at the press of a button, so minimizing the need for speech. Even voice transmissions can be translated into digitalized blips that, reflected by the satellite, allow rapid and undistorted two-way contact regardless of distance. The cockpit will become a quieter and less busy environment. Flight managers at base will be able to monitor much of the data now displayed on the pilot's instrument panel.

The International Civil Aviation Organization's committee on Future Air Navigational Systems (FANS) says that it expects the satellite system to be ready for the North Atlantic routes 'possibly by 1995', and in worldwide service by the year 2010. The 18 satellites already in use (April 1989) will be increased to 42 and provide total global navigation. A successful test flight, guided entirely by satellite, has flown from Stansted, London, to Madrid.

The satellite programme is well on its way but, in the longer term, progress towards supersonic and hypersonic space travel is less sure. There are plenty of plans for super-Concorde designs – faster, bigger and longer-range – but most of them are still on the drawing-board.

If futurology is a science, it is a short-sighted one. Like weather forecasting, long-range predictions become more of an art form that, however inspiring, loses much practical validity. At the same time it would be wrong to discount the visionaries who have played such a prominent part in aviation – R. J. Mitchell's Spitfire and Sir Frank Whittle's jet engine were both mocked and delayed by the practical men of their day. Then, as now, the rate of progress was governed by the tension between what is theoretically possible and what is politically acceptable.

Nowhere is this more evident than in the prospect of supersonic flight for all. Although the first barrier has been breached by Concorde, the next leap forward hangs fire for the same old reasons –although they appear in a new guise. The conventional wisdom is that within the next two decades it will be technically feasible to girdle the earth in a hypersonic passenger airliner in about an hour and a half. Today's subsonic airliners must fly below the speed of sound or, as it is called, Mach 1. (The division between supersonic and hypersonic flight comes at Mach 5.) Speeds could rise twelvefold

221

from Concorde's Mach 2.2 to Mach 25 – roughly from 1,400 to 17,000 miles an hour. This is the consensus that emerges from a confusing number of projects now under way in the United States. They also promise less breath-taking designs for progressively faster aircraft with speeds of Mach 4 and upwards in the years leading up to the ultimate goal of Mach 25.

As will be seen later in the wider context, it is immensely significant that the chief obstacle lying in the path of progress is not the lack of technical know-how, but the universal fear of harming the environment. This is forcing the industry into technical detours to satisfy political rather than aerodynamic imperatives. The lesson of Concorde has been learnt. In the short term, these call for a quieter engine to reduce airport noise. Designers are at work on a variable cycle jet that will behave as quietly as normal airliners during take-off and landing, before switching over to the noisier and faster thrust needed to propel the plane at supersonic and hypersonic cruising heights.

Rather less has been said about the elimination of the sonic boom caused by breaking the sound barrier – an unmet need that has hagridden the career of Concorde. Much aerodynamic research is being done, we are told, towards reducing the notorious bang 'to socially acceptable levels'.

France, West Germany and Britain are working on European high-speed designs. The promise of British Aerospace's HOTOL project (horizontal take off and landing) has been dashed by the withdrawal of government support after it had won acclamation for its advanced engine designs but, as an unmanned craft, any civil applications were a long way ahead. Politics have also dogged many US research projects, most notably work on the National Aerospace Plane (NASP) 'X-30' which has brought new materials to resist the highest speeds and advanced aerodynamic theory needed for hypersonic flight. But its shape remains conceptual. Congress is questioning the need for it so that its immediate future lies in the balance. Government attitudes are ever fickle.

Addressing the nation on the future US space programme in 1986, President Reagan held out some mighty promises about air travel: 'Yes, we're going forward with research on a new Orient Express that could by the end of the next decade take off from Dulles airport, accelerate up to twenty-five times the speed of sound, attaining low earth orbit and flying to Tokyo within two hours.' Three years later this sounds like a piece of high-flying rhetoric considering that we still lack the technology for a Mach 4 aircraft, much less Mach 25 and while the X-30 project lies in the doldrums.

The third hurdle that now faces the designers comes from the new concern about pollutants that are damaging the earth's protective ozone layer. Among these are the oxides of nitrogen emitted by the exhausts of high-flying jet planes, and a NASA chief has admitted that this problem will be 'extremely difficult' to overcome. General Electric, one of the designers of advanced engines, agrees that the critical problem of building an acceptable supersonic transport engine is its effect on ozone depletion in the upper atmosphere.

In the face of these three purely enviromental injunctions, it looks as if any further prognostications about how they can be met are best left to the experts. Until now, aviation has had a free run largely unhindered by the social considerations and the constraints on growth that affect other industries. That era appears to be ending. The significant change is that we who are living down below are now setting the specifications to be met in the air and it appears that this new factor – partly born out of the Concorde programme – is about to shape the future of air transport more than any other. The signs are already with us.

If, as has been said, every 'now' is a collision between the past and the future, we need to look more closely at the impact of the events that are taking place under our noses today and which throw their shadows ahead.

The human limits to growth

Concern about the environment is not the only manifestation of customer power that is beginning to shape events. By October 1989, the package holiday trade was planning to sell 3 million fewer holidays in the following season and anticipated a drop in demand of 20 per cent. Remembering that leisure travel as a whole is said to account for 80 per cent of all airline traffic, the move is significant. British package holidays, mostly to Mediterranean sunspots, nearly trebled in the ten years from 1978. The travel industry may hope that this is a short-term trend but, with some reason, there are people who see this as whistling into the wind to mask a shaken confidence in the old rate of growth. By the winter of 1989, the leading firm of Thomas Cook was predicting that 1,000 agencies could fold within a year.

It would be rash to extrapolate this into a permanent tendency from a time-base of only a couple of seasons, but when the likely causes of the shrinkage are reviewed, they appear on a broad enough front to suggest a sea-change in attitudes to holiday travel. As it will be argued later, this may be no bad thing for safety as well as for other wider considerations.

Tourism responds to economic forces like any other trade and lower disposable incomes in a period of high interest rates must have their effect. Yet the tour operators themselves are evidently aware of a changing demand that is by-passing the economic trends of the moment. A growing number of holiday advertisements now promote resorts with promises of unspoilt features, clean beaches, quiet and secluded surroundings (coupled with more generous compensation for delays and mishaps). Fewer people seem to be seeking the delights of the mass and brash resorts of old. There is talk of a switch to longer-hop holidays of higher quality and price – but fewer of them.

While a new disenchantment with today's air travel is evident enough, the part played by each of a number of disincentives is less certain. The more obvious of these include hassle and delays at the airports – a nationwide blight in the United States since the de-regulation of the airlines and an ordeal compounded by anti-terrorist security checks at international terminals. In Europe, unfilled hotels at resorts may reveal a growing ennui with the package holiday circuit around the endlessly replicated concrete blocks that are now so similar the world over, coupled with a new awareness about the polluted coasts of the Mediterranean and beyond. These, and a sense of having done it all before, have tarnished the glamour of air travel.

There are signs too that people at home and abroad are beginning to question the benefits of the huge mass exchange of tourists every season. Neither are the real benefits to national economies as great as governments would have us believe. Although tourism is a prime earner of foreign currency, in Britain the balance sheet of incoming and outgoing currencies shows a historic deficit. Only two years in the period from 1980 to 1988 have shown a credit balance (1988 returned an overall loss of over £2 billion).

Seasonal tourists complain that cities and high spots are jam-packed with others of their kind and on returning they find their own home purlieus invaded to capacity by the counter-waves of incoming tourists. Disquiet about the tourist blight, once a fashionable middle-brow topic, has filtered through to the popular media with enough effect to disturb the politicians. The policy of open-ended growth is no longer sacred and the new awareness of its consequences suggests that the belief in foreign travel as a pleasure has taken a knock. Sea-side resorts at home are enjoying a new appeal. There is a ring of truth in the maxim uttered by the American Kenneth Boulding, a self-styled 'econoclast', which declares that anyone who believes that exponential growth can go on for ever in a finite world is either a madman or an economist. Following the thought that many a truth is spoken in jest, there is a serious element here that makes a point about the air travel industry as a whole.

Whether the future of tourism depends most on family bank balances or as much on human inclinations, may be a matter for argument. But there is one instinct that defies the economists, yet concerns us most here – the fear quotient. As discussed earlier, surveys suggest that about one passenger in four is seriously worried by the prospect of flying, although this can be no more than a broad measure of a volatile response. Each major aircrash adds to their number for a while: in February 1989, after a series of crashes – including the Pan Am jumbo at Lockerbie, Scotland, and the British Midland 'motorway' disaster 18 days later – British tour operators shed 2.5 million Mediterranean holidays as a result of the fall-off in bookings. One headline ran 'Fear of flying kills 2.5 million holidays', and a leading operator commented that there was 'no doubt that some passengers were becoming concerned about flying'. An airline, which declined to be named, said that it had had many calls from first-time or infrequent flyers inquiring about the type of aircraft they might be flying in and for assurances that they had undergone the proper maintenance procedures.

The following month a *Time* magazine survey (13 March) asked respondents if they thought it safer now to fly on commercial airplanes than it had been five years ago: 64 per cent thought it was less safe, 20 per cent said it was safer and 11 per cent saw no change.

Ironing out the temporary effect of periodic disasters or clusters of crashes, there seems small doubt that a comprehensive survey would show the trend of fear to be rising. One conjectures that threats of

225

terrorism and hi-jacking – elements that are measurably on the increase – play at least an equal part to the risk of accidents. The prospect is a sombre one and efforts to 'take out' the terrorist from the air have not yet been rewarded. Should we not be asking ourselves not only who committed the act of terrorism, but examining much more closely why it was done? While political solutions may be elusive, they lie at the base of the threat and an inch won on this front is worth a yard of security checks.

Leaving aside the threat from terrorism, it was suggested that a more discerning attitude to air travel might bring other benefits. There is a new realism in the travel trade that has sensed a call for less cut-price hassle and higher quality – a call that has accompanied the demise of a number of operators and airlines. The free-for-all that followed de-regulation in the United States has allowed the bigger fish to swallow smaller competitors and so gain monopoly over local captive markets, resulting in shoddy domestic services and many unco-ordinated route structures. Gathering concern about a lowering of safety standards has become a public issue. The take-over fever caused by the struggle for survival between carriers has led to massive borrowings. In October 1989 US Transport Secretary Samuel Skinner issued a public warning that these huge debts could put airline safety at risk. Market pressures to pay down principal and to meet the interest payments could, be said, 'threaten a carrier's ability to meet its obligations including fleet replacement, aircraft repair and maintenance, security, crew training and other safety-related expenses'. As these effects of total deregulation begin the pinch at both ends of the industry, one hopes that the present scene may now contain the seeds of its own salvation.

It is evident, too, that more people know more about the ways of the industry than ever before: the covers are being lifted not only on the operators (for example, the safety of twin-engined planes crossing the Atlantic), but also on safety standards and accident causation. We learn as we fly.

If we are looking for a solution to the present distempers of air travel, this surely points to the heart of them. The call for openness, which runs through much of this book, follows the recognition that this is the surest route not only to making flying safer, but also towards an industry that marches in step with changing social aspirations. Change may be too modest a word to describe the

226

worldwide swing towards the protection of the environment – a sudden conversion that is laying down new limits to so many human activities. The magnitude of the coming changes has hardly dented the popular imagination.

Speaking in the House of Commons in November 1989, the chairman of its select committee on energy, Sir Ian Lloyd, declared that the threats to the global climate demanded a worldwide programme by comparison with which all previous endeavours, even in war and in putting a man on the moon, 'will pale into insignificance'. He foresaw that in the USA, the worst-case scenario meant that virtually the whole of its Gross National Product would be absorbed in protecting its coastal regions. Fitting aviation into this picture, it must mean that any plans for future power plants and propellants that are not demonstrably user-friendly, are destined for the archives.

On a less elevated level, the boom days of holiday hype and an over-extended industry may be over. The tourist trade hotly deny any such trend but – to coin a phrase – they would, wouldn't they? A more discriminating public may bring the benefits of limited growth and a healthier aviation industry, backed by less crowded skies and airports. Collision avoidance systems, satellite navigation, anti-fire cabin spray equipment are still years away from general operational use. These, and many other branches of safety technology now in the pipe-line, would stand a better chance of keeping pace with growth than they have done in the past years of breathtaking expansion in the industry's operational wing.

These growing pains have been evident in many accidents. Remember, for instance, the defective design of the DC-10 cargo door that led to the Paris crash in 1974: the model was rushed out to beat the launch of Lockheed's competing TriStar. A slogan erected in the DC-10 manufacturing plant at Long Beach, California, boasted 'Fly before they roll'. The lethal failure stemmed from the primitive and scamped design of the cargo door.

The government men

The British seem to have a propensity for disarming their competence. It would be hard to find any airline pilot who does not agree that the air traffic controllers at London's Heathrow airport and

West Drayton (the area control centre) are a model of excellence throughout the world. The few lapses that have occurred have been shown to be mainly due to under-staffing, overwork and stress, coupled with delays in providing controllers with the tools they need – higher capacity radar and other electronic systems to handle the rapid growth in traffic.

The UK Air Accidents Investigation Branch likewise sets the highest professional standards that have won worldwide renown. But, in comparison with its American counterpart, its statutory remit severely limits its functions and its potential. British professionalism and the integrity of its public services have few rivals, but while on the one hand having created this noble asset, we proceed to clip its wings with the other. Chapter 8 laments the AAIB's legal duty to avoid the apportionment of blame on those involved in accidents, while the US National Transportation Board suffers no such inhibition. This leads to the contention that the British approach contributes less to air safety than it could.

In practice, the contrast between the two is reflected by the form of their respective accident reports. The AAIB's findings, technically unimpeachable, are humbly addressed by the Chief Inspector of Accidents to the Secretary of State for Transport and put on sale to the public at rather high cost. NTSB reports, signed not by the technical men but by the Chairman and his board, are addressed to the world at large. While they may make recommendations to the Federal Aviation Administration, the NTSB is free to dig deeply into the human background: for example, it quotes interviews with pilots, colleagues, relatives and other witnesses without inhibition or apparent fear of the consequences.

These contrasts have led to a call for an autonomous UK safety agency on the pattern of the NTSB which in the USA ranks as a government agency and is independent of other state departments. Wider powers would enable it to monitor, chivvy and bring to public account any sector of the industry concerned: to apportion blame when and where it is due, to suggest remedies, research and advice to government, freed from any other departmental duties and the normal inhibitions that are inseparable from a responsibility to individual ministers. Thus enlarged, it would absorb some of the functions of the Civil Aviation Authority and, as an independent force, carry a punch equal to that of its American counterpart – a

228

power that both the AAIB and the travelling public certainly deserve.

The distance between the sense of public accountability shown by air safety officials in the United States and the conduct of their counterparts in other countries has already been examined. The American commitment to full and immediate disclosure and – where it is called for – the allocation of blame, is important enough in the debate about air safety to highlight a more recent and vivid example.

On the night of Wednesday 20 September 1989, 2 women were killed and 59 people were rescued from the water of the East river after a USAir Boeing 737–400 had tried to abort its take-off but over-ran the runway at New York's La Guardia airport. It had accelerated to 161 m.p.h. – the point of no return – when its captain decided to abandon the attempt and brake to a stop, but the plane overshot the water-bound end of the runway and broke up into three sections.

Two days later the deputy chairman of the National Transportation Board spoke to the media from the site of the crash to deliver a detailed run-down of the accident in a frank and critical manner, although it was understood by all that his preliminary comments were subject to the findings of the full and final inquiry into the accident. He noted that the co-pilot, who had been at the controls, was flying a B 737-400 for the first time in his career: he had mistakenly pressed the auto-throttle disengage button, to be corrected by the captain who then pressed the correct engage button. On the take-off run, the plane failed to maintain the centre-line of the runway and drifted to the left: it continued to slew left until its track was midway between the centre and the hard edge of the runway, whereupon the captain took control and ordered the take-off to be abandoned. (Later investigation pointed to a number of factors, both human and mechanical, that contributed to the crash.)

The media read much into the fact that, not unusually, neither of the pilots was available for interview by accident officials for 36 hours after the crash: although flightcrew are encouraged to talk to investigators immediately, there is no firm legal requirement for them to do so. The growing threats of criminal prosecutions and civil compensation suits have led pilots and their advisers to prepare their case before going on record.

229

This kind of misunderstanding by the media is part of the price paid for the policy of full disclosure in the United States. The fact that the NTSB chief went public as soon as he had gathered the material facts – uninhibited by the final results of the official inquiry which, as everybody knows, would follow in due course – speaks of a real commitment to the flying public and a recognition of their right to know. (His words even reached British radio listeners early on the Saturday after the crash: it corrected doubts about the integrity of the B 737-400 which had been aired in the press in the wake of the USAir crash, wrongly linking it with the British Midland 'motorway' disaster to the same type of plane in January 1989.)

As for the airlines and others directly affected in the industry, the knowledge that their conduct will become publicly known while attention is still focussed on the disaster, is a much more effective sanction than a written report published in the margins of the media months or years after the event when memories of the event have faded.

It is sadly inconceivable that this exemplary frankness would be shown today in other state accident inquiries – whether it be Britain, France, Spain or elsewhere.

Gamekeeper or fox?

At present, lip service is given to the need for an independent agency by placing the AAIB under the Department of Transport, while the Civil Aviation Authority has an ambiguous and hybrid identity: although not a government department, it manages to have a vicarious voice in Parliament. These wafer-thin walls that are supposed to isolate the three seats of activity – AAIB, CAA and Parliament – are too artificial to prevent the voice of the executive from echoing through from one to another. Probably the nearest model for an independent safety agency is the UK's Office of Fair Trading which is enabled to stand at a greater distance from government departments than are lesser breeds of quangos.

As mentioned in Chapter 11, the status of the CAA exhibits another ambiguity from the double role imposed upon it. Charged with the duty of promoting the interests of the industry it serves, it must also act as its policeman – two roles that appear to be mutually exclusive. Even given the impartiality of a Solomon, it would be hard to reconcile these duties without compromising one at the expense of the other.

Down the line, this takes the form of the appointment from the ranks of employees in the industry of men charged with a CAA supervisory role – a cost-saving practice that is shared (and as much criticized) in the US FAA. The adverse consequences of these delegations upon safety are evident from a number of accident histories told in earlier chapters. It is an odd logic that expects an agency to protect our safety with one hand and then lets it put the fox in charge of the hencoop with the other.

Under the present dispensation, the CAA is also burdened with the quasi-commercial function of paying its way by charging for the services it provides. Considering that the Authority's primary function is to serve an industry whose life-blood is tourism – the nation's second largest source of trading income – a method of book-keeping that ignores these large benefits to the public purse seems to show a political rather than a logical principle at work.

Conclusions

Those who have read thus far will not find a last conclusion surprising. If the completed jig-saw reveals any unity of theme, it is that the many forms of secrecy in human affairs too often rebound to punish us. Fudge, fear, hokum and hype stand as the foremost enemies of safety in the air.

But there will be some who will not share my convictions. Critics of the system must expect challenges and be ready to meet them. The part played by the press and television in air safety is likely to be one of them.

The excesses of the gutter press, or what might be called the unacceptable face of journalism, has fed a deep distrust of the media that scorns its motives and its self-appointed role as tribune of the people. Many people in the aviation world are implacably opposed to its instant judgments, its intrusions and often its political mendacity, all of which they see as a hindrance to official accident investigations and a meddlesomeness that does a disservice to air safety.

Admitted. But it can be equally irresponsible for these sceptics to ignore the positive value of a system that permits free comment and, as it were, to throw the baby out with the bath-water. Philosophical opinions about the difference between liberty and licence do not take us much further here because they are entirely subjective. There can be bad reporters (or editorial directors) just as there can be bad professionals in the industry. As has been said, the fact that the media is the liveliest watchdog is due less to its own merits than to the decrepitude of the offical pack. Unlike the achievements of the last century, the pace of change has overtaken the capacity for constitutional reform with the result that the phalanx of officialdom – headed by Parliament and the courts – has become ever more remote, inaccessible and costly to the individual with a problem. Untroubled by change, bureaucracy has spun its web of secrecy and delay around the centres of power. Government agencies are not likely to be first in the field of inquiry into any misdeeds, mistakes, scandals or iniquities.

It is the media men, the paparazzi, who are usually first on the scene, followed by the official troops, impelled into action by leaks and rumours that can no longer be ignored. The politicians and their officials may resent being up-staged in this way. Media motives, they retort, are suspect and campaigning journalists guilty of a 'hypocrisy of concern' or they adopt a radical chic posture to mask their true aims of higher circulation and profits. They may be told that such sensationalized and alarmist intrusions into public calamities amount to no less than the heinous trade of disasterology.

Some of this is well-earned abuse and all that can be said is that those whom the cap fits should wear it. But it cannot invalidate the process of investigation itself. The integrity of an inquiry or report must be left for the public to judge for itself. By analogy, a medical diagnosis may be good, bad or indifferent but the process itself is the first step to a cure. The aetiology of air accidents can be no different.

As a pejorative term, disasterology seems to extend to journalism the idea of ambulance-chasing lawyers. As an exercise in hindsight wrapped in righteous indignation, it is of course a cheap and nasty sham if, as is often the case, it sets out with the aim of finding fault or pillorying a scapegoat. It can be the kind of magisterial comment on a disaster that comes under the threadbare caption 'an accident waiting to happen'. At the other end of the scale, there are those who share my sense of overwhelming frustration at the limited amount of truth that breaks the secrecy barriers in the aftermath of an accident.

Only the few accidents that raise a major political row are likely to reach a thorough public inquiry or a full-scale appeal in the courts, generally several years after the event. The Pan Am bomb disaster at Lockerbie and the sea ferry sinking at Zeebrugge are examples. The remainder get cursory and often ill-informed media attention, followed by a late official report that is stale enough to pass by largely unnoticed. The exceptions, in which the press alone took a campaign to the bitter end – such as the Thalidomide scandal and the Turkish DC-10 crash near Paris – belong to the past and are now unfashionable in that quarter. To an encouraging extent, however, the tradition of investigative journalism still lives on in television and sound broadcasting.

In contrast to America, the narrow remit of routine offiical inquiries in Britain ensures that they do not trespass into the political or social arenas. As has been said, the apportionment of blame, as an example, would be to enter forbidden territory. It is left to the higher echelons of officialdom to pose remedies for executive action, to identify trends, and to draft briefs for their political superiors. But these deliberations are closely protected from view by the twin bulwarks of secrecy and delay.

If one were to choose examples to support this view, the first might be the continued use of highly flammable plastics in the cabins of airliners and the deaths from asphyxia and fire that these have caused over the past two decades (Chapter 14). Remedies have so far been marginal and achieved at an excruciatingly slow pace. Public awareness of the dangers would have halted their use long ago, if it had been known that an airliner fuselage is an infinitely more dangerous place to inhabit than a house or cinema or other place of public assembly.

The second award for delay and obfuscation would go to the kind of bureaucratic hocus pocus that emerged before and after the first crash of a Boeing 747 at Nairobi in 1974 (pp. 118–27). The snag in its warning systems, fatal on this occasion, had caused near-accidents before and was well documented. It is lamentable that the bitter comment made after Nairobi – that there has to be blood spilt on the runway before the regulators take notice – reads like an epitaph to so many air disasters.

The common strand, then, that runs through a study of accident causation shows itself to be secrecy in one or other of its various forms. If that case has been made, it follows that relentless inquiry is the first necessity: eternal paranoia, it could be said, is the price of safety. Public pressure may eventually bring a cure for the British malady, but in the present climate it is probably wise to remain sceptical on that score. If that is a gloomy conclusion, it is cheering to borrow a thought from air pilot lore: pessimists are happy, optimists are dead.

Those who remain doubtful about the prime role of the media as an instrument of change have a lot of history to contend with. One does not need to look far to see that the first targets of the conventional political coup or rebellion are the broadcasting stations and the control of the press. At a more civilized level, the two Watergate newspaper reporters whose attentions led to the dethroning of a president achieved more than his political opponents could do, and television coverage of the Vietnam war that brought its horrors into the American homes hastened its end. In Britain, a formidable number of major scandals, some of immense public concern, have been detected by the media rather than any other agency. The legal limits of fair comment in Britain, already more stringent than in most western countries, have been drawn tighter around the media by recent decisions in the courts. Pre-publication injunctions and the recent strengthening of the official secrecy laws are among the tighter government controls that have been reviewed earlier.

Yet new curbs seem about to be imposed. The call for a new statutory right of privacy is backed by a majority of opinion, fuelled largely by the understandable outrage at the worst excesses of the tabloid press. Opponents of the measure, including thè English Bar Council, among others, believe that it will tighten the gag on the proper exercise of free speech too far. There are indeed reasons why a right of privacy could turn out to be a double-edged sword. It could be a godsend to the crooks and fraudsters who have evaded the police, but a hindrance to the television teams and reporters who have been bringing them to book with remarkable regularity. Many bigger fish – city magnates, arms and drug dealers and even spies – might still be in high office or living in comfortable obscurity if they had been armed with a right of privacy as a shield against meddlesome reporters.

Anger at the gutter press is a healthy reaction, but it is an emotive response that needs to be balanced by an understanding of the virtues

234

as well as the vices of free comment. To that extent, a law of privacy in an already secretive society could take us backwards rather than forwards.

War and peace

The reasons for the British attitude to state censorship and secrecy may be diverse, but it is a persuasive thought that the legacy of two world wars comes high on the list. During the First World War the mounting struggle for survival, unprecedented and unforeseen, brought Draconian powers to control the news on both the military and home fronts, and with it an unquestioning obedience to authority. The Second World War revived the machinery of censorship and, to my mind, the presumption in favour of state secrecy bred under these wartime regimes dies hard. It lingers on like a wartime scar that the nation has borne with it into peacetime.

The control of the skies, for so long a theatre of war, still carries some of the aura of a restricted territory in which civilians are granted only a concessionary interest. For example, unlike the intrusions of roads and railways, air law deprives them of any formal right of complaint or compensation for aircraft noise or passage over private property (a statutory exception to the rights of ownership which would otherwise include the airspace above it). Civil aviation, of course, would be impossible without the abrogation of some private rights, yet it is remarkable that, until very recently, its *carte blanche* in the skies has hardly come under question. There are other minor symptoms that reveal the same quasi-military presumptions of secrecy: much operational data is effectively restricted to airline staff or pilots, and it even remains an offence for 'unauthorized persons' to eavesdrop on the radio channels used by aircraft – a publicly funded service.

However far wars have nourished these secretive habits in all branches of officialdom, it is encouraging to see that secrecy, as a national disability, is at last attracting so much public debate in the press and broadcasting. Without the media as a platform, there would be small hope of checking the inroads of censorship.

The air is an unforgiving element. However safe flying may become, by its nature the consequences of fault or failure will remain more dire than in most other phases of modern life. Regrettable

though secrecy is in public affairs, its intrusion into air safety is an
iniquity.

Appendices

Appendix I Survivors' guide

Chapters 14 and 15 considered crash survivability, particularly escape from cabin fires and the forthcoming improvements based on trials of the SAVE water-spray system, which are some years ahead. Safety regulations require trials that demonstrate that all passengers must be able to evacuate the cabin within 90 seconds, but in too many accidents this limit has been exceeded in practice and many lives have been lost as a result. To a great extent, survival depends on the behaviour of the passengers in an emergency. But they need to know just what to do. There are a number of things that can better our chances of survival in a crash, and the following expert guide comes from Squadron Leader Mark Lewis of the Royal Australian Air Force who produced it while on the staff of HQ Air Force Inspection and Safety Center, Norton Air Force Base, California, USA.

He points out that often one person survives but the passenger in the next seat does not. Heading a list of factors that explain that difference, he cites the sheer will to survive as the most important. So fatalists beware. A person who is pre-armed with the determination to beat the dangers will anticipate them – and that is half the battle. Here are the main points he offers as a survivor's guide.

- Plan to get yourself out – crew members may not reach or even see you: they may be occupied elsewhere or themselves disabled.
- Pre-planning helps. Survival can depend on dress: long sleeves and long pants are good. Wool is protective, synthetics are bad. The more layers the better. Very tight or very loose clothes are equally dangerous. High heels are dangerous – take them off.
- Sit next to an exit if you can. Look at the doors as you go in – which way do the handles turn, does the door push in or out? Many fire accidents have ended with piles of victims pressed against doors they could not operate. You may be fighting in the dark, so count the rows to your chosen exit.

- Listen to the safety briefing and identify the exits. Fat or elderly people next to the exits may block your escape route and may be incapable of working the doors (they can weigh 60 lbs and need healthy muscle to move). So choose the most hopeful exit.
- Most people understand the pre-crash brace position mimed by the cabin crew before the flight, but legs are not usually mentioned. Keep them out in front rather than tucked under your seat, or they may swing forward and hit the metal bar under the seat in front. Broken legs don't speed escape. Many victims have died this way.
- Sit down and slide on the escape chute – don't jump. It's a long drop from the wing and you can't run clear with a broken leg – again, a common fate.
- Without smoke-hoods, a water-soaked cloth held over nose and mouth will partially filter the smoke and toxic gas, and a dry cloth is better than nothing at all. Take shallow breaths and as few as possible – hold it if you can until clear of smoke and fumes.
- If you're not injured, don't wait for help. Help from aircrew is a bonus, but you are one amongst hundreds. If you are with your family, discuss your plans with them in advance. In smoke, have them hold on to your hand or belt. With small children and infants, think through how you will handle them.
- If you are prepared, you may become the survivor that helps others to safety. So *don't* be a fatalist.

Appendix II Lufthansa World Accident Survey, March 1989

The several sources of accident statistics referred to in this book have their own validity in time and place of origin, but by their nature recent figures are most relevant to present and future events. Captain Heino Caesar, General Manager, Flight Operations and Safety Pilot of Lufthansa German Airlines, permits me to select and summarize some of the findings that bear on the topics I have discussed. Much of his work is the result of personal research and is based on a computer data base containing all relevant accidents to large jet aircraft since 1959. Sabotage, hijack and war-like actions are excluded. The survey was contained in his address to the Airline Insurance Conference in London. Besides being up-to-date, both its source and the venue add to its value. I have extracted points solely in the order of their importance to my own text.

- 'In a very generalized way' among the causal factors for total losses, cockpit crew error accounted for 76 per cent.
- Flight safety trends, expressed by geographical area, show the most endangered regions to be South America followed by Africa and the Far East. The safest areas are Australasia followed by North America and Europe.
- Small airlines have more accidents, proportionately, than those with 50 aircraft or more. 'It is alarming' that small companies that in total own 2 per cent of all aircraft, produced nearly 40 per cent of all accidents.
- Seventy-five per cent of all accidents occur within the vicinity of an airport (this accords with earlier figures). Airlines are suffering nearly half of their total losses in 2 per cent of their flight time, that is, at landing and take-off.
- Long-range heavy, multi-engined airliners have a relatively bad safety record – 2.4 times worse than that of short- or medium-range aircraft. (It is suggested that, since most accidents occur at landing or take-off, this points to lack of experience due to less frequent stops on long-haul routes.) Older narrow-bodied airliners show a worse record than new wide-body types.
- While some other sectors of flight have become safer, landing losses have increased: this is put down to airport deficiencies.
- From 1959 to 1988 (inclusive) there were 495 total losses of big jets (including, in this case only, 49 due to sabotage/war). Of these, 392 were operational airline 'accidents', in which 15,198 passengers and 1,471 crew died. The annual average over the last 30 years is 507 passengers and 49 crew killed.
- Over the last five years, the average insurance value of a commercial jet was 22 million US dollars: with an average of 19 jets lost each year, annual losses total $418 million.
- Costs for each person killed are more alarming, now standing at nearly one million dollars in the United States (1988).
- 'We must consider a loss rate of 10–15 heavy jets with an average of 400–800 lives lost per year.'

Photograph credits

Index

242

243

245

246

A Selected List of Titles Available from Minerva

While every effort is made to keep prices low, it is sometimes necessary to increase prices at short notice. Mandarin Paperbacks reserves the right to show new retail prices on covers which may differ from those previously advertised in the text or elsewhere.

The prices shown below were correct at the time of going to press.

Fiction

☐	7493 9026 3	**I Pass Like Night**	Jonathan Ames	£3.99 BX
☐	7493 9006 9	**The Tidewater Tales**	John Bath	£4.99 BX
☐	7493 9004 2	**A Casual Brutality**	Neil Blessondath	£4.50 BX
☐	7493 9028 2	**Interior**	Justin Cartwright	£3.99 BC
☐	7493 9002 6	**No Telephone to Heaven**	Michelle Cliff	£3.99 BX
☐	7493 9028 X	**Not Not While the Giro**	James Kelman	£4.50 BX
☐	7493 9011 5	**Parable of the Blind**	Gert Hofmann	£3.99 BC
☐	7493 9010 7	**The Inventor**	Jakov Lind	£3.99 BC
☐	7493 9003 4	**Fall of the Imam**	Nawal El Saadewi	£3.99 BC

Non-Fiction

☐	7493 9012 3	**Days in the Life**	Jonathon Green	£4.99 BC
☐	7493 9019 0	**In Search of J D Salinger**	Ian Hamilton	£4.99 BX
☐	7493 9023 9	**Stealing from a Deep Place**	Brian Hall	£3.99 BX
☐	7493 9005 0	**The Orton Diaries**	John Lahr	£5.99 BC
☐	7493 9014 X	**Nora**	Brenda Maddox	£6.99 BC

All these books are available at your bookshop or newsagent, or can be ordered direct from the publisher. Just tick the titles you want and fill in the form below. Available in:
BX: British Commonwealth excluding Canada
BC: British Commonwealth including Canada

Mandarin Paperbacks, Cash Sales Department, PO Box 11, Falmouth, Cornwall TR10 9EN.

Please send cheque or postal order, no currency, for purchase price quoted and allow the following for postage and packing:

UK — 80p for the first book, 20p for each additional book ordered to a maximum charge of £2.00.

BFPO — 80p for the first book, 20p for each additional book.

Overseas including Eire — £1.50 for the first book, £1.00 for the second and 30p for each additional book thereafter.

NAME (Block letters) ....

ADDRESS ....

....

....